Tacit and Explicit Knowledge

Tacit and Explicit Knowledge

HARRY COLLINS

The University of Chicago Press
Chicago and London

The University of Chicago Press, Chicago 60637
The University of Chicago Press, Ltd., London

Paperback edition published 2013
Printed in the United States of America

22 21 20 19 18 17 16 15 14 13   3 4 5 6

ISBN-13: 978-0-226-11380-7 (cloth)
ISBN-13: 978-0-226-00421-1 (paper)
ISBN-10: 0-226-11380-9 (cloth)
ISBN-10: 0-226-00421-X (paper)

Library of Congress Cataloging-in-Publication Data

Collins, H. M. (Harry M.), 1943–
Tacit and explicit knowledge / Harry Collins.
p. cm.
Includes bibliographical references and index.
ISBN-13: 978-0-226-11380-7 (cloth)
ISBN-10: 0-226-11380-9 (cloth)
1. Tacit knowledge. 2. Knowledge, Theory of. I. Title.
BF317.5.C65 2010
001—dc22

2009026465

⊗ This paper meets the requirements of ANSI/NISO Z39.48-1992
(Permanence of Paper).

For Susan

CONTENTS

PREFACE

This book began as an attempt to put my earlier studies of tacit knowledge together in a coherent way. I thought it would be easy, but I soon found that rather than having my arms around the whole subject, my grip was precarious. I am not the only one who thinks the existing literature on tacit knowledge is less than clear. The confusions are found in all the disciplines that take tacit knowledge to be part of their concern, including philosophy, psychology, sociology, artificial intelligence, economics, and management. This book is, first, an attempt to resolve these confusions and, second, with the resolution in hand, an attempt to produce the coherent account of tacit knowledge. It can also be seen as a foundation for the tacit knowledge–based Periodic Table of Expertises Robert Evans and I set out in *Rethinking Expertise* (2007) and as a setting for the more detailed analysis of the notion of polimorphic and mimeomorphic actions found in my and Martin Kusch's *The Shape of Actions* (1998). Thus, this book amounts to the completion of a three-book project to analyze knowledge from "top to bottom"—from the nature of expertise to the nature of actions, with the nature of tacit knowledge in the conceptual middle.

Polimorphic actions are actions that can only be executed successfully by a person who understands the social context. Copying the visible behavior that is the counterpart of an observed action is unlikely to reproduce the action unless it is a mimeomorphic action, because in the case of polimorphic actions, the right behavioral instantiation will change with context. Here it will be concluded that, for now and the foreseeable future, polimorphic actions—and only polimorphic actions—remain outside the domain of the explicable, whichever of the four possible ways "explicable" is defined. This has significance for the success of different kinds of machine and for the way we teach. If we are ever to make the tacit knowledge associated with polimorphic actions explicit, such that we could build machines that can mimic polimorphic actions, then what I will call "the socialization problem" will have to be solved first.

The argument set out here begins with the claim that existing treatments of tacit knowledge are unclear about what is meant by the terms "tacit" and "explicit." It is also argued that while it is true that all explicit knowledge rests on tacit knowledge, we would have no concept of the tacit without the explicit. The second argument is that existing work fails to separate three phenomena, all known as tacit knowledge, which are quite different and which I refer to as weak, medium, and strong tacit knowledge. These have to do, respectively, with the contingencies of social life (relational tacit knowledge), the nature of the human body and brain (somatic tacit knowledge), and the nature of human society (collective tacit knowledge)—RTK, STK, and CTK.[1] It is CTK that requires a solution to the socialization problem if it is to be explicated. The experience of the individual who is learning something new usually involves elements of all three—though not necessarily in sequence—and the resulting "Three Phase Model," I suggest, is more fundamental and general in its reach than previous approaches. The experience of the individual, however, unless examined with analytic determination, is pretty much the same whichever of the three types of tacit knowledge is being encountered, and acquiring all of the types is often part of the same learning experience; that is why existing analyses work reasonably well when they address narrow problems and why it has not been noticed that very different things are being talked about. It is, nevertheless, vital to separate these different kinds of tacit knowledge if mistakes are to be avoided when the gaze is lifted and more ambitious problems and projects are addressed.

Some of the components of this book have been discussed before. The distinction developed in chapters 5 and 6—the difference between the body and the collectivity—were to some extent worked out in my contribution titled "What Is Tacit Knowledge?" that was included in *The Practice Turn in Contemporary Theory* (2001) and in an article published in *Organization Studies*, "Bicycling on the Moon" (2007). However, a complete classification of tacit knowledge emerged only with the idea of relational tacit knowledge, which is new to this book and arrived only with the most enormous

1. I originally called relational tacit knowledge "contingent tacit knowledge" but it later occurred that it might be useful to have distinct acronyms for the three types. "Relational" captures the idea that whether these pieces of knowledge are tacit or made explicit depends on the relation between the parties. The other two types of tacit knowledge do not become explicit when social arrangements change.

struggle. Sometimes, the simplest things are the hardest to see if one starts from the wrong position, and I now see that my *Artificial Experts* (1990) has relational tacit knowledge mixed up with other kinds of tacit knowledge. A good few of the examples used here have also been used before in *Artificial Experts, The Shape of Actions,* and other books and papers. The old examples remain well suited to make the points, and there are many new examples, too. It is only in this book that I have begun to understand exactly how they all fit together, and that is one of the main aims of the exercise.

ACKNOWLEDGMENTS

More in the way of a division of labor than usual went into this book. It would not have been written had it not been for the extended discussions of tacit knowledge I had with Rodrigo Ribeiro during my supervision of his PhD. Many of the early problems and tentative attempts at solutions arose as we talked things over and the results have frequently found their way into the book. Later this discussion spread to the weekly seminar of the Centre for the Study of Knowledge, Expertise, and Science, and exactly who contributed what to the earliest formulations would be hard to say. On the other hand, only as the manuscript's twenty-six drafts unfolded over a couple of years did the nagging logic of the page bring out the vital importance of starting with the explicit and finding a place for relational tacit knowledge, thus driving the argument to coherence while indicating the proper roles and meanings of the elements.

In the course of writing I frequently e-mailed authors and asked them to send this piece or that and they invariably responded with grace and more advice than I had any right to expect. I will make no attempt to list everyone who has helped in this way, because I would be bound to miss some. Nevertheless, I must thank Stephen Gourlay and Edgar Whitley for keeping me in touch with existing studies—each sending me a reading list. Being a terrible scholar myself, I contrived to get them to do some of my work for me, but the end result is my responsibility, as is the fact that there is no attempt in the book to do the kind of review of the literature that would give proper recognition to everyone who deserves it. I thank Terry Threadgold for convincing me, one way or another, that the apparatus of semiotics was not what I needed for the opening chapters and it would be better to start afresh from more elemental components. Rob Evans, Stephen Gourlay, Martin Kusch, Trevor Pinch, Rodrigo Ribeiro, Evan Selinger, and Edgar Whitley read the manuscript and sent me really useful feedback. And Chicago's two anonymous readers did a wonderful job, perfectly understanding where the book was coming from, basing their criticisms on where it was trying to go, and doing their best to help it get there. Nevertheless, all mistakes and infelicities that remain are my responsibility. Finally, Christie Henry, my editor, has been brilliant—a full partner in the enterprise. So has Mary Gehl, my copy editor, and, on the basis of my past experience, so too will Stephanie Hlywak be, when it comes to putting the thing out into the wider world. If you are ever lucky enough to get University of Chicago Press interested in a project of yours, bite their arm off.

INTRODUCTION

The Idea of Tacit Knowledge depends on Explicit Knowledge!

We can know more than we can tell.

—Michael Polanyi, *The Tacit Dimension*

Now we see tacit knowledge opposed to explicit knowledge; but these two are not sharply divided. While tacit knowledge can be possessed by itself, explicit knowledge must rely on being tacitly understood and applied. Hence all knowledge is either tacit or rooted in tacit knowledge. A wholly explicit knowledge is unthinkable.

—Michael Polanyi, "The Logic of Tacit Inference"

The Territory of Tacit Knowledge

Tacit knowledge is knowledge that is not explicated. In this book, tacit knowledge will be analyzed and classified by reference to what stops it being explicated; there are three major reasons why tacit knowledge is not explicated; therefore, there are three major types of tacit knowledge. Of course, if we are going to say why things cannot be explicated, we first have to understand what is meant by "explicated." That gives rise to the structure of this book: explain "explicit," then classify tacit.

Tacit knowledge drives language, science, education, management, sport, bicycle riding, art, and our relationship to machines. That is to say, tacit knowledge makes speakers fluent, lets scientists understand each other, is the crucial part of what teachers teach, makes bureaucratic life seem ordered, comprises the skill in most sports and other physical activities, puts the smile on the face of the *Mona Lisa*, and, because we users bring the tacit knowledge to the interaction, turns computers from *idiots savants* into useful assistants. The aim of the book is to reconstruct the idea of tacit knowledge from first principles so that the concept's disparate domains have a common conceptual language. To switch the metaphor, the idea is to generate a Google Earth–type view of the entire united domain that will make it possible to "zoom in" on any area with ease and understand its relationship with all the other areas. The case studies and analytic discussions of tacit knowledge that we already have in hand—the bike riding, the laser building, the sapphire quality-measuring, the car driving, the natural language speaking, the breadmaking, the transfer of knowledge between organizations, and so forth—will turn out to be aspects of the same territory seen from different vantage points. With the new map, we will see where those known bits of the territory are separated by mountains, where they are linked by passes, and where it was always just a matter of level ground.

Tacit knowledge currently lives a varied life in a range of academic disciplines, including philosophy, psychology, sociology, management, and economics; and by right, it ought to play a large part in the world of artificial intelligence. Those who first think of the term as associated with Michael Polanyi are likely to go straight to his famous example of bicycle riding: we can know how to ride a bicycle without being able to tell anyone the rules for riding, and we seem to learn to ride without being given any of the rules in an explicit way—our knowledge of the ability to ride a bike is tacit. This book will have a lot to say about the bicycle example, as it is one of the sources of confusion about the meaning of tacit knowledge, confounding knowledge embodied in the human body and brain—*somatic tacit knowledge*—with knowledge "embodied" in society—*collective tacit knowledge*.

Philosophers of one kind will associate the idea with Wittgenstein's argument that rules of action do not contain the rules for their application—the rules "regress." Thus, to apply a rule like "do not walk too close to others in the street," one must know what "too close" means and how it varies from circumstance to circumstance, and one must know another set of rules to know how to recognize what kind of circumstance it is, and so forth. Given that we cannot produce an exhaustive list of such rules, this must mean that when we live our lives according to them we must be using tacit knowledge to know how they are to be applied. Philosophers of another kind will associate the idea of tacit knowledge much more with the human body and its relationship with the world of practices as discussed by, among others, Heidegger and Merleau-Ponty. In this book this conception will be discussed by examining Hubert Dreyfus's application of these ideas and will be shown to be just one conception of tacit knowledge—somatic tacit knowledge.

Developmental psychologists, insofar as they use the term "tacit knowledge," are also likely to think of it as having to do with the body. It is a fact that children nearly always learn the conceptual structure of the world through their body's interaction with the environment. Furthermore, our language turns on the makeup of our bodies—had we no knees we would have no notion of "chair"; had we no fingers we might have a different counting system; had we no eyes our conceptual world might be very different.

Sociologists of knowledge might have encountered the notion through my case studies of the way scientists learn to repeat laboratory manipulations, such as building working lasers or making delicate measurements. To accomplish these things requires enough personal contact between the scientists to enable things that are not spoken to be passed on in ways that may not be visible or apparent. Sociologists such as myself have highlighted what is here called collective tacit knowledge—which is located in society. In later chapters, my own early studies will be criticized for not paying enough attention to the different types of tacit knowledge and the different ways of passing them on that were ready to be examined if I had thought about it.

Those who come at the problem from the management literature might well take Nonaka and Takeuchi's discussion of the bread-making machine as their paradigm: Nonaka and Takeuchi describe the way the previously tacit knowledge associated with kneading dough for bread was elicited and formulated so that it could be reproduced in mechanical form in a bread-making machine. Again, the book will show that Nonaka and Takeuchi's conception is very narrow. They think the notion of tacit knowledge is exhausted by knowledge that just happens not to have been explicated but could be given a bit more effort. Nonaka and Takeuchi are dealing only with *relational tacit knowledge*. To understand the bread maker properly, a lot more is needed; a way of analyzing the bread maker more fully is offered in appendix 1.

Economists, or those whose concern is with "knowledge management" at the level of the organization, might think about tacit knowledge in terms of strategies for capturing elusive skills by recruiting people who already have needed tacit knowledge or by acquiring whole businesses that already have capacities embedded in their personnel that the existing firm lacks.

The three-way classification of tacit knowledge—relational, somatic, and collective—is the basis of the new map, but it will not be encountered until chapters 4, 5, and 6. Chapters 1, 2, and 3 are more of a philosophical ground-clearing exercise. The tacit cannot be understood without first understanding the explicit, and these chapters are an exploration of what "explicit" means. Readers who want to understand tacit knowledge in order to be able to use the idea effectively in their practice can skim the first three chapters—or at least jump over any parts that might seem overelaborated. For those readers who feel they won't understand the explicit or the tacit until they have worked through the relationship between digital and analogue strings and other such esoterica, the first three chapters might be worth a critical examination. A lot of it is a matter of stating the obvious—but stating the obvious is not always easy when one begins with a confused domain. Regardless, all readers will want at some stage to capture the sense of table 4 (p. 81) which offers four meanings of "explicable"; the later chapters refer back to these four meanings.

Problems with the Term "Tacit"

The problems of the existing discourse of tacit knowledge begin to show themselves as soon as one looks at the term itself. Thus, the *Chambers Dictionary* defines "tacit" as "unspoken" or "understood or implied without being expressed directly." But Polanyi talked of *can* and *cannot*: "we *can* know more than we *can* tell." In the dictionary definition, "tacit" is descriptive—tacit knowledge is knowledge that is not explicit—but in Polanyi's usage, "tacit" is knowledge that *cannot* be made explicit. The tension between "is not" and "cannot" permeates the entire discussion. Consider the antonyms: the opposite of the dictionary definition of tacit is "explicit"; the opposite of Polanyi's definition is "explicable." It is bound to be confusing if, to turn the thing on its head, two different parts of speech—"explicit" and "explicable"—have the same antonym—"tacit." And this is not to mention the fact that "explicable" generally means "can be explained," whereas the opposite of "tacit" means plain and clear and expressed directly; the first is about knowledge, the second is about style.

Questions about the Use of "Tell"

Moving from the term "tacit knowledge" to Polanyi's phrase "we can know more than we can tell": what is "know," what is "tell," and what are "can" and "cannot?" Consider these questions:

If I encrypt the map reference for a U-boat rendezvous, do I know more than I can tell and is the encrypted message explicit or tacit knowledge? Before the Rosetta Stone was translated, could its contents be told? And how did I know that it contained knowledge of any kind and wasn't just a pretty pattern? What if a mathematical ignoramus is told some rules for solving differential equations that he or she cannot use?—Is this telling or is it not telling? What if I overhear a few remarks exchanged between two people that I can't understand, but, noticing my puzzlement, they explain at length what they were talking about?—Is that tacit knowledge being converted to explicit knowledge? And what is meant by "understand" in this context? What if I give my love a single red rose? Is this telling her something? If I don't know what is on a computer CD, but I find out when I place it in the drive and the computer fires up, has something tacit been made explicit? Have the programmers told a pocket calculator how to do arithmetic? What if I use the record-and-playback method to train a machine to spray chairs with paint? Have I told the machine something explicit or does the machine now have the tacit knowledge of the trainer? Does a sieve have the knowledge to sort big items from small items, and, if "yes," did the designer "tell" the sieve how to do it? Does my cat have tacit knowledge of how to hunt? It doesn't have explicit knowledge! What if I have a special grip on my golf club that ensures that my hand assumes the right position? Is the special grip telling my hand what to do? What if I can write out the mechanical formula for balancing on a bike? Does that mean that bike riding is explicable? What if I tell a novice that he or she should look well ahead, not down at the ground, when trying to learn to ride a bike? Is that telling the novice how to ride explicitly? If I act for reasons that are subconscious, are they tacit and do they become explicit if the psychiatrist uncovers them?

In chapter 3 these questions will be answered and by the end of that chapter such questions should no longer seem puzzling. By the end of the book, questions about whether all or some tacit knowledge can ever be made explicit should no longer seem puzzling either, or, at least not so puzzling as they are now.

Coming to realize that something that initially seemed clear is confusing can be a perverse kind of progress. With the concept of tacit knowledge there are two important sources for this backhanded rejoicing. The first is the long-running attempt to build "intelligent" computers. Once upon a time, the only candidate for the meaning for "know" in Polanyi's "we can know more than we can tell" was human knowing. What was explicit was explicit to humans. Now we have to think about whether or not something should count as "explicit" if some set of instructions will enable a machine to carry out a task: should a successful computer program count as explicit knowledge even if a human could not execute it? When it was a matter of, say, print, it was easy: if a human could not use it in some direct way then it was not explicit. But now machines can use or transform written symbols that humans can't and there is a new problem about what explicit means. It seems to me that some past discussions, not least my own, have involved a degree of sliding around this issue. The mistake is to believe that understanding human experience is the route to understanding knowledge. Rather, to understand human experience one must start by trying to understand all the things that might count as knowledge and then work out how humans might use them. The growth of automation has provided new problems and more demanding questions about what knowledge might be even though it remains the case that, in the last resort, humans are the only knowers.[1]

The second source of perverse progress has been the new understanding of the social setting of scientific knowledge that was first developed in the 1970s. This has advanced our understanding of human knowledge in general. In particular it has given us a much deeper understanding of the meaning and implications of Polanyi's claim that "all knowledge is *either tacit* or *rooted in tacit knowledge*."[2] The studies of science that began in the 1970s revealed that even the paradigm of explicit knowledge—scientific data or the algebraic expressions of theory—can be understood only against a background of tacit knowledge. This has revealed that the idea of the explicit is much more complicated than was once believed. In appendix 2, some specific new understandings that have come out of these two broad developments are listed and briefly described.

The claim that explicit knowledge depends on the tacit is, however, all too easy to overread. I have been earnestly assured by scholars that there is no such thing as explicit knowledge—it is all tacit. But if all knowledge is tacit, what is it that is "rooted in tacit knowledge"? Polanyi's very formulation shows that a distinction between tacit and explicit has to be preserved, though it doesn't show us exactly where the distinction lies or how it works.

1. When I say that one must start by thinking about all the things that might count as knowledge, I do not mean to claim that anything like classical epistemology is being pursued. First, for the sociologist of knowledge, or the Wittgensteinian philosopher, there is no classical epistemology—knowledge cannot be found in the absence of the activities of humans. The point is that we must *start* with an attempt to think about knowledge in a way that goes beyond human experience if we are to understand that experience properly. The starting point is to think of knowledge as "stuff" that might also be found in animals, trees, and sieves and then try to work out from this starting point what it is that humans have. Human experience alone is too blunt an instrument for the task.

2. Polanyi 1966, 195 (original emphasis).

The Tacit depends on the Explicit!

What the mistaken claim that all knowledge is tacit does indicate is that, mostly, explicit knowledge is harder to understand than tacit knowledge. Most writing on tacit knowledge takes it to be the other way around. Though the tension between tacit and explicit goes back at least as far as the Greeks, it was modernism in general and the computer revolution in particular that made the explicit seem easy and the tacit seem obscure. But nearly the entire history of the universe, and that includes the parts played by animals and the first humans, consists of things going along quite nicely without anyone *telling* anything to anything or anyone.[3] There is, then, nothing strange about things being done but not being told—it is normal life. What is strange is that anything *can* be told.

Once one sees how normal and natural it is to do things without being able to tell how they are done, one also sees that a good part of the notion that there is something strange about *tacit* knowledge is parasitical on the idea of explicit knowledge. If "all knowledge is *either tacit* or *rooted in tacit knowledge*," the explicit seems to be parasitical on the tacit—which it is to the extent that the explicit is without significance in the absence of the tacit. But the reverse is true when we consider not the knowledge itself but our *idea* of the tacit. The idea of the tacit is parasitical on the *idea* of the explicit. The idea that the tacit was special could not occur to anyone until explicability came to be taken to be the ordinary state of affairs and that moment was a recent one in human history, and one that is fast drawing to a close. Thus, in the traditional discussion—if something that has only been going on in full flow since the middle of the twentieth century can be called "traditional"—the idea of the tacit seems hard only because, mistakenly, the explicit has been taken to be easy. The pioneers of the idea of tacit knowledge, reacting to the enthusiasm for science and computing typical of the 1940s and '50s that made the explication of everything seem easy—no more than a technical problem on its way to being solved—had to fight to create space for the tacit, and, as a result, they made it into something mysterious.

It is time to redraw the map. I will argue that many of the classic treatments of tacit knowledge—those that have to do with bodily skills or the way the human brain works in harmony with the body—put the emphasis in the wrong place. What the individual human body and human brain do is not much different from what cats, dogs, and, for that matter, trees

3. Gourlay (2004) points out that the same applies to the socialization of children.

and clouds have always done. While humans encounter bodily abilities as strange and difficult because we continually fail in our attempts to explicate them, there is nothing mysterious about the knowledge itself. It is knowledge that, *in principle,* can be understood and explicated (in one sense of table 4) by the methods of scientific analysis. In practice it may be hard to describe the entire picture but it is hard to develop a complete scientific explanation of many things. In spite of the possibility of scientific explanation in principle, it remains true that for most individuals, if not all, that the body is central to the acquisition of knowledge. This, however, says less about the *nature of knowledge* than has been assumed; what it does indicate is something about the nature of human beings and how they acquire knowledge. More profoundly, it also remains true that the nature of the body does, to a good extent, provide the conceptual structure of our lives, but that conceptual structure is located at the collective level, not the individual. One of the main projects of this book is to demote the body and promote society in the understanding of the nature of knowledge.

There is a second reason the discussion of tacit knowledge is parasitic on explicit knowledge: the need to transmit knowledge from person to person. We want to know the most efficient ways to get people to be competent at doing new things. The cheapest and easiest way to enhance peoples' abilities is to tell them things. You can tell people things by giving them books to read or sending them messages over the Internet or, at worst, sitting them in classrooms and talking at them. But these methods will not work unless the thing that is to be transferred can be transferred via a medium such as print or talk. If it cannot be thus transferred, the process of raising the level of peoples' abilities is going to be the much harder, longer, and more expensive process of socialization, or apprenticeship, or coaching, or the equivalent—all of which require that everyone be physically shifted into the same geographical space and in fairly small numbers. Print or talk, if it works, can transfer abilities from one to many—it can be "broadcast"; apprenticeship cannot. Likewise, building machines that can do things for us is often said to depend on "making the tacit explicit." So more often than not, questions about the nature of tacit knowledge are tied up with questions about the transfer of tacit knowledge, and questions about the transfer of tacit knowledge are tied up with questions about converting the one type of knowledge into the other.

At the risk of being accused of political incorrectness, I'll sum up everything that has been said so far with the punch line of an old joke about a lost visitor asking a passerby for directions to Dublin: "I wouldn't start from here." As will become clear in chapter 1, we are going to take this advice and start from somewhere else.

What Will Be Found in the Chapters

Chapter 1 is an attempt to approach the notion of explicit knowledge anew from the most reduced set of elements. There are "strings" and there are "entities" (humans, animals, and inanimate objects). "Strings," as I define them here, are bits of stuff inscribed with patterns: they might be bits of air with patterns of sound waves, or bits of paper with writing, or bits of the seashore with marks made by waves, or irregular clouds, or patterns of mould, or almost anything.[4] Sometimes these strings have no effect on the things they impact, sometimes, being physical objects, they have a causal or mechanical effect, and sometimes they are "interpreted" and their effect comes from the meaning imputed to the patterns. "Explicit knowledge transfer" involves communication via strings of the ability to accomplish new tasks. Strings are the building blocks of what semiotics refers to as signs, symbols, and icons; strings, however, do not begin with the freight of inherent meaning that makes the notion of signs, symbols, and icons so complicated. On the one hand, the semiotic terms connote meaning; whereas on the other hand, whether they are actually read as having meaning or not is context dependent. The term "string" is more basic: a string is just a physical object and it is immediately clear that whether it has any effect and what kind of effect this might be is entirely a matter of what happens to it.

Consider this analogy: imagine I pick up a stone and throw it at a coconut, which then falls. I label the stone and similar stones "knocker downers." Then I need a whole philosophy to explain the puzzling behavior of knocker downers: it is inconsistent—sometimes they knock things down and sometimes they don't. It is the equivalent philosophical puzzle that can be avoided if one starts with strings, not signs.

There is a crucial distinction between "strings" and "languages." A language is a set of meanings located in a society, whereas, to repeat, strings

4. This has nothing to do with "string theory" as in physics. The metaphor in string theory is "lengths of string," whereas the metaphor used here is, as found in *Chambers Dictionary*, "a set of things threaded together or arranged as if threaded." In some ways, it is akin to the usage in computing—an ordered set of symbols in one dimension—but is more general still, including, as it does, the physical medium on which the information contained in the pattern is expressed.

are just physical objects. A condition for the existence of languages is some kind of approximate representation of meaning by strings; strings are the means by which languages are shared and there can be no language without sharing. Unfortunately, because language and strings are so intimately related, they are sometimes confused. But the strings are not the language. The difference between strings and languages is more sharply defined in chapter 1, which begins by looking at all the ways that strings can interact with other things.

If one is concerned with the transmission of knowledge between humans, one must be concerned, willy-nilly, with what is fixed. If the quintessential question is "How does A learn from B?" Whether B wants to build a laser, bake bread, speak sentences, live in society, or whatever, then the fact that what B learns is mostly not exactly what A intended, or the fact that the meanings of "bread" and "laser" are social constructs, perhaps with political significance, are not to the point. What is to the point is that something with a relatively fixed meaning that carries a degree of technical empowerment has to be transferred. Thus, in spite of the fact that translation can rarely be done without loss or transformation, this is not what is emphasized here. This book emphasizes that which is *not* lost in translation.[5]

This approach, then, is in *tension*, with most of what has gone on in the broad area of science and technology studies and semiotics over the last three decades or so, but it is not in *opposition* to it. Rather, a new kind of question is being asked of the same materials—instead of stressing the flexibility of interpretation, attention is turned to the fixedness. Everything that has been discovered during these decades about the degree of indeterminacy in the interpretation of a string remains true and a central, and a still unresolved puzzle, is how there can be any fixedness at all. In chapter 2 the puzzle is "papered over" with the term "affordance."

Chapter 2 also analyzes the nature of strings much more carefully, exploring the distinction between analogue and digital strings and showing how strings are continuous with cause and effect as it is ordinarily encountered in the world.

Chapter 3 reconstructs the everyday use of the notion of the explicit in terms of the basic elements. It takes up what has been worked out in chapters 1 and 2 and uses it to show how to resolve problems in the ordinary talk of explicable knowledge. Ordinary (academic talk) about these issues

5. For a treatment with the stress on the positive transformations associated with translation, see Latour 2005.

is reconstructed. The second part of chapter 3 comprises answers to the list of questions set out on pages 4–5 above.

In chapters 4, 5, and 6 the notion of tacit knowledge as it is encountered in everyday life is split into the three parts that have already been mentioned and which can also be thought of as weak, medium, and strong tacit knowledge.

Chapter 4 deals with weak, or relational, tacit knowledge, which is knowledge that is tacit for reasons that are not philosophically profound but have to do with the relations between people that arise out the nature of social life. The reasons range from deliberate secrecy to failure to appreciate someone else's need to know. A characteristic of weak tacit knowledge is that, in principle, with enough effort, any piece of it could be rendered explicit. That not all of it can be rendered explicit at any one time has to do with logistics and the way societies are organized.

Chapter 5 looks at knowledge that is tacit because of the way it is inscribed in the material of body and brain—this is somatic tacit knowledge. *Somatic-limit* tacit knowledge is knowledge that can be written out (at least in principle) but cannot be used by humans because of the limits of their bodies. In general, machines of the right design can execute somatic limit tacit knowledge. *Somatic-affordance* tacit knowledge is knowledge that humans can execute only because of the affordances of the substance of which they are made. In general, machines cannot execute somatic-affordance tacit knowledge, because they are not made of the right kind of materials. This chapter departs from many modern treatments of the problem of tacit knowledge by insisting that somatic tacit knowledge is not a difficult problem for philosophy or practice and that a heavy diet of examples of tacit knowledge that turn on the abilities of the body has not been helpful.

Chapter 6 looks at strong, or collective, tacit knowledge—the knowledge that the individual can acquire only by being embedded in society. This is called "strong," because we know of no way to describe it or to make machines that can possess or even mimic it. Strong tacit knowledge is a property of society rather than the individual. The individual, being a parasite on society, can learn it, however. The individual can learn the practices and the language, or, as has been argued and will be reiterated here (under the label "interactional expertise"), can learn the language without learning much in the way of the practices. The ability to draw upon strong tacit knowledge—to be a parasite on the body of the social—renders humans unique: neither animals nor things can do it. This idea is labeled "Social Cartesianism."

Part III of the book, comprising chapters 7 and 8, looks backward and forward. Chapter 7 uses the new analysis to look back briefly at existing discussions of tacit knowledge found in economics, management, and philosophy; it also re-examines some of my previous studies and explains how they might have been improved.

Chapter 8 sets out the promised map, based on the Three Phase Model of tacit knowledge, which is unfolded in chapters 4, 5, and 6. It shows how the tacit and the explicit relate in our lives and suggests how the map might be used when thinking about how to explicate tacit knowledge for new applications and when such an exercise is likely and unlikely to succeed.

PART I

Explicit Knowledge

ONE

Strings and Things

> It shows great folly . . . to suppose that one can transmit or acquire clear and certain knowledge of an art through the medium of writing, or that written words can do more than remind the reader of what he already knows on any given subject. . . . The fact is, Phaedrus, that writing involves a similar disadvantage to painting. The productions of painting look like living beings, but if you ask questions they maintain a solemn silence. The same holds true of written words; you might suppose that they understand what they are saying, but if you ask them what they mean by anything they simply return the same answer over and over again. Besides, once a thing is committed to writing it circulates equally among those who understand the subject and those who have no business with it; a writing cannot distinguish between suitable and unsuitable readers. And if it is ill-treated or unfairly abused it always needs its parent to come to its rescue; it is quite incapable of defending or helping itself. (Socrates, in Plato, *Phaedrus*, 275d5–275e5)

Strings

When the explicit is discussed, it is usually taken as having to do with humans communicating via signs, icons, codes, or some such. As intimated in the introduction, here the analysis starts at a more basic level—interaction between physical objects, sometimes referred to as strings and sometimes as entities. At the outset, humans, animals, and other objects are treated as undifferentiated entities. The aim is to pull them apart, but the differences are to be established—not assumed. Another unusual feature of the analysis is that very little attention is paid to transmitting entities; nearly all the work of analysis concerns strings and their impact on things with the producers of strings being part of the background.

Strings, as explained, are bits of stuff inscribed with patterns. Strings in themselves are not meaningful. A string is simply anything that is neither random nor featureless.[1] I said earlier that there were "strings" and "entities," but strings are entities and entities are strings, so what a string is and what an entity is depends on what is going on at the time.

Technically, a string always contains "information" in the sense connoted by "information theory."[2] Information content is a physical feature of a string that refers to the number and arrangement of its elements. Strings, though they always contain information, are not always used to transmit information.

It will be useful from time to time to speak of the "elements" in the patterns pertaining to strings. Elements include the 0s and 1s of binary code, the letters of the alphabet, features of the patterns of smoke clouds, electrical states in silicon microcircuits, the positions of ratchets and cog wheels, the repeated shapes on patterned wallpaper, the notes made by a songbird, the atoms or molecules in a substance, or the features of lichen on a rock, and so forth.

The terms "string" and "strings" mean the same thing, since a long string is also a set of strings. Furthermore, *elements* of a string can also be strings in themselves and vice versa. For example, the letter "A" is *an element* in strings made up of letters of the Roman alphabet, but it is also *a string* when expressed in binary code as "01000001." And to illustrate the point made immediately above, "01000001" is both a string and a set of strings. Here the terms "string," "strings," and "elements" will often be used to refer to the same thing—after all, strings are just entities. In the actual analysis, the meaning of terms should emerge from the context without difficulty.

Interaction of Strings and Entities

Strings, in our manner of speaking, sometimes interact with entities. They can affect entities in four ways. (1) A string is a physical thing, so it can have a physical impact. (2) A string is a pattern, so it can impress, print,

1. The idea of "signs" and the like gives too much ontological priority to the class of "strings that are interpreted." We just don't know how to divide the world up between, say, signs and non-signs, because you cannot tell which is which by looking at them; that which is a sign (like that which is a "knocker downer") is relative to its producer and reader and this makes such terms complicated and confounding in use.

2. Usually associated with Claude Shannon (e.g., Shannon 1948).

Table 1. How strings affect entities.

1 PHYSICAL IMPACT	2 INSCRIPTION	3 and 4 COMMUNICATION	
		3 Mechanical	4 *Interpreted*

or "inscribe" a similar pattern on an entity in many different ways. (3 and 4) A string can change an entity in a more fundamental way than mere inscription—it can cause it to do something or give it the ability to do new things that it could not do before. This is called "communicating," and it can be done in two ways: (3) a string can communicate "mechanically," as when a new piece of code is fed into a computer or a human reacts to a sound in a reflex-like way; and (4) a string can communicate by being interpreted as meaningful by a human. Table 1 sets out the four ways strings affect entities.

The way a string impacts on an entity depends on the relationship between them. If the string is physically hard, it will more easily have a physical affect on an entity on which it might impact; if the entity is soft, the effect of the impact will be greater. Likewise, the way inscriptions work and the type of inscription that results from the impact of a string depend on the physical instantiation of string and entity. "Ink" can be used to enhance the effect. Whether the impact of a string results in a communication also depends on the string and the entity. The analysis will often proceed through exemplification of how differences in string and entity produce differences in outcome. For example, what will be called condition 4 and condition 5 of communication depend on different kinds of changes in the entity enabling what might be a mere impact or an inscription to become a communication.

In table 1, category 4 is emphasized because it is special. In the other cells any kind of entity can be found, but category 4 refers only to humans. Humans are also found in category 3, however, and sometimes the two categories have been confused. An example of the difference is the sergeant major's shout of "'Shun!" to soldiers on the parade ground. If the troops are thoroughly trained, they will respond to the shout in a reflex-like way. We do not say the sergeant major "told" the troops to come to attention; we say he "brought the troops to attention," implying a mechanism at work. But before the troops learned to respond in this mechanical way they would almost certainly have gone through a stage where they responded to the same shout in an interpreted way. They would have heard the "'Shun!" and interpreted it as the command "come to attention," and would have moved their bodies in response.[3] Novice troops will hear themselves being *told* to come to attention. Thus, if the sergeant major assembles novice troops and tells them that, for a change, when he says "'Shun!" he wants them to relax, they will probably be able to do it. But when they are well enough trained to be causally affected by "'Shun!" they will not hear the command as an instance of "telling"; under these circumstances it would be very hard to change their normal reaction to "'Shun!" from coming to attention to relaxing. The idea of basic training is, of course, to make soldiers respond to orders on the battlefield without needing the time for conscious processing that could create the opportunity for fear.

Discussions of explicit knowledge generally concern themselves only with interpretation (column 4 of table 1) and that is why terms such as sign, icon, and code play such a central role. Unfortunately, the exclusive use of terms like these, which connote meaning and interpretation, encourages the mechanical impact of strings on entities to be confused with the interpreted impact. Thus computers are especially well represented among the entities found in the mechanical effect category (column 3 of table 1), but they are often thought to occupy the interpretation box, too. One of the advantages of starting at the basic level of strings is that this kind of confusion is less likely to occur. The three top-level categories set out in table 1 are now described.

Strings and Physical Impact

A string is always a physical thing. For example, the Bible is a string; at one time the recommended treatment for getting rid of a "ganglion"—a swelling in a tendon sheath—was to smash down on it with the family Bible. In the case of this use of a string, its information content and the fact that it is a string as well as a physical thing is not relevant.

More significantly, a string cannot have an effect that involves inscription or communication unless it also has an initial causal effect on the entity. Thus, in the course of reading this book (let us say), you have encountered a microdot containing a potentially interpretable string hidden within one of the full stops. But if this was the case it has had no causal ef-

3. This is not necessarily the case—one could teach the appropriate response to "'Shun!" to troops who did not know the language through a regime of punishment and reward as one might train an animal.

fect on you and therefore you have not been affected by it—not even to the extent of being inscribed with its pattern. If the microdot were magnified so that you could see the string clearly then it might inscribe or communicate something. Or again, if the sergeant major whispered the command "'shun" rather than shouted it, it would not have the initial causal impact required for subsequent causal or interpreted effects to follow—sergeants major have to shout loud if they want their strings to have mechanical or interpreted effects. As we will see below, when machines are described as "reading codes," they are transforming strings into a form that can have an appropriate causal impact on humans which, in turn, allows for the possibility of interpretation.

Strings and Inscription

A printing press contains strings in the form of the letters on plates. When these plates are introduced to the paper they produce an inscription. The same applies to my computer and its printer, except that the strings are in the form of electrical signals in the computer, which undergo various *transformations* prior to doing whatever is necessary to produce the paper inscription. Any forceful impact by one physical object on another is likely to cause an inscription.

Inscriptions may be temporary or lasting. When I speak, I temporarily inscribe the air with vibrations which are transformations of the movements of my larynx. Thomas Edison invented a way to transform the vibrations in the air into more permanent strings.

Humans can be inscribed with strings in various ways. For example, I might write a message on your forehead or pin a note to the back of your jacket without you noticing. More interesting is memorization and recitation; the multiplication table is a noteworthy example. Schoolchildren used to learn to recite the table in "parrot fashion." When I was young I had such a table inscribed in my person; decades on, it is still instantly accessible by chanting. Thus, if I want to know the result of multiplying 7 by 8, I simply chant the table and the result "seven eights are fifty-six" appears; it is just as though I had it written backwards on my forehead and I looked in the mirror. In this example, my "knowledge" of the multiplication table is exactly the same as it would be if I had not memorized the tables myself but had a slave who did not speak my language but who had been taught to memorize and recite the tables by a regimen of punishment and reward. A flick of the whip and the seven times table would be recited for my use without

Box 1. Knowing words without understanding (the case of the deaf).

"They'd write on the board, and we would copy it. Then they would give you good marks and you would swagger about. But what did those words mean? Ha! Nothing! It all went past us . . . . Yet those two-faced people would give us good marks and pat us on the head." (P. Ladd, *Understanding Deaf Culture: In Search of Deafhood* [Clevedon: Multilingual Matters Ltd., 2003], 305)

the slave needing to interpret it in any way whatsoever—the slave would be simply *inscribed* with the table.[4]

Sometimes inscription is mistaken for proper teaching. A poignant example is illustrated in box 1, where a deaf boy describes his misery at having to learn meaningless words—meaningless because, without being immersed in the bath of speech from an early age, deaf children have difficulty in acquiring native spoken languages. We can assume that in the case of the deaf as represented in the box, the teachers believed they were accomplishing more than they were.

Memories are laid down in chemical and electrical pathways in that mysterious organ called the brain, so it is tempting to think that even the memories that make for mere recitation lead a somewhat more exotic metaphysical life than printed symbols. But the exotic appearance is due only to the fact that the medium on which the symbols are inscribed is three-dimensional human tissue rather than paper or silicon and that the means of "writing" is somewhat less well-understood in this case than when it is produced by a pen, typewriter or computer. Learning the multiplication table "by rote" is, however, just inscription, and it is metaphysically continuous with printing.

Strings and Communication

Taking the cue from Wittgenstein's "ask for the use, not the meaning," whether the transfer of a string counts as a communication depends on the outcome. Here a communication is defined as follows: *A communication takes place when an entity, P, is made to do something or comes to be able to*

4. Searle's famous Chinese Room (see chapter 6, below) is simply inscribed with language and cannot work as well as many readers of the example take it to work (see Searle 1980 and, for the critique, Collins 1990, esp. ch. 14.)

do something that it could not do before as a result of the transfer of a string. If the string is merely inscribed on an entity, "Alpha"—imagine it written on the forehead—then Alpha can enable a second entity, "Beta," to read the string merely by letting Beta, say, look at the inscribed forehead. But this is not what we mean here by "coming to be able to do something." On the other hand, entity Alpha will have been enabled to do something if it can use what has been transferred in some productive way—if it results in Alpha having new and useful knowledge. For example, if I can use the answer "56," which I obtain by reading the inscription on my slave, in another sum, or to affect a purchase, or some such, then the transfer of the string to me resulted in a communication as well as an inscription. A marginal case is when the entity is enabled to do something like answer a question in the game of Trivial Pursuit but is not able to do anything more with the new "knowledge." That kind of case just has to be kept in mind and will appear again.[5]

An act of communication is like jumping across a gap between two buildings. Sometimes the gap is narrow enough to be jumped without anything special needing to be done. But very often, communication fails because the gap, as first encountered, is too wide to jump. In a subset of the difficult cases, something can be done to improve the jumping ability or to narrow the gap. One can sometimes increase the success rate by modifying the string so it can jump further or one can improve the chances by building out from the far side so there is less far to jump. There are two ways of making each kind of adjustment. Thus, including the fact that the gap might be narrow enough not to cause problems in the first place, there are five enabling conditions of communication. In this section the first four are described, while the fifth will be explained a little later.

Condition 1

Condition 1 is when everything is already in place so that the gap can be jumped and a string interacts with an entity and the result is a trouble-free communication. In condition 1, nothing has to be done to "enable" communication. An example is when a computer responds to the typed command 10 × 2.54 and produces the result 25.4 or a human is asked to multiply 10 by 2.54 and produces the result 25.4. The more interesting conditions are 2, 3, 4, and 5, because the communication fails in the first instance and something has to be changed to enable the gap to be jumped.

5. In Collins and Evans 2007, this kind of case is called "beer-mat knowledge."

Condition 2.

Condition 2 has already been mentioned, but it is included here for completeness. It is a physical transformation of a string that enables it to have the causal impact on an entity which is the precondition for communication. The example that has already been discussed is magnifying a microdot. As it stands the microdot cannot "jump the gap," because it has no causal impact on the entity; when magnified, however, it can have a causal impact. For continuity, let us run through the condition as it applies to the sum 10×2.54. Imagine that the keystrokes required to get the computer to do this sum are written out on a piece of paper. For the computer to be able to actually do the sum—for the gap to be jumped—the paper string will first have to be transformed into a string that can apply physical pressure to the appropriate keys—a string of finger positions, say. This is a simple string transformation.

Condition 3

In condition 3, a short string fails to result in a communication, but a longer string succeeds in jumping the gap. For example, an early computer fails to respond to 10×2.54 because it had no calculator program; it can succeed when the command is supplemented with a calculator program (an additional string making a longer string when the two strings are taken together). Another example would be a computer fed a program written in the language C++ when it has no means of handling code in anything other than BASIC. If, in addition, if it was fed the longer string comprising the instructions for handling C++ it would then be able to handle the initial string. In the same way a human may not be able to do the sum until the short string, "multiply 2.54 by 10" is enhanced with the string: "to multiply by ten, move the decimal point one place to the right."[6]

A nice illustration of condition 3 communications is the following old joke (which, coincidentally, happens to be about old jokes):

> A man walks into a strange pub and notices that every now and again one of the locals shouts out a number and all the other locals laugh. The landlord, seeing the puzzled expression on the face of the visitor explains that the locals have been telling the same funny stories to each other for so long that they now just refer to them by number. Instead of going to all the trouble of

6. Those familiar with the game Awkward Student (Collins 1985) will know that there is no *guarantee* that the calculation will eventually be done correctly, however long the string. Nevertheless, longer strings sometimes work when shorter strings do not.

> reciting an entire story one of the locals will just call out a number and the others laugh because the number brings the story immediately to mind. The visitor tries it and calls out some numbers but each effort is met with silence. The landlord, feeling sorry for the visitor, explains that the crucial thing is not the joke itself but the way that you tell it.

The punch line is not important. The moral is that an entire story can be represented by a number but that the number will only be understood by a "local"—that is, someone to whom the number affords the story. But each joke could be communicated to the visitor if a longer string was used—the string in which the joke was originally transmitted. The joke is not so different to things that happen in real life: for example long-standing members of bureaucratic organizations sometimes talk almost entirely in acronyms that outsiders cannot understand.[7]

These examples show why the idea that anything tacit can be made explicit so long as the strings are long enough is so seductive. It seems from these examples as thought it is just a matter of making the strings longer. But just because condition 3 sometimes works it does not follow that it will always work.[8]

Condition 4

Sometimes the transmission of even a longer string will not result in a communication. Perhaps this is because no one or no thing has the wit or the will to create the longer string that will do the job or perhaps because no string, however long, will work. The question of whether long-enough strings always could be made to do the work of failed short strings lies at the center of the whole idea of tacit knowledge. Irrespective of the answer to this question, in some cases where short or longer strings will do not the job, a fixed change in the physical form of the entity will enable the strings to succeed in jumping the gap where they did not initially succeed. This is condition 4.

7. I have had such an experience as an outside advisor to a Whitehall committee.

8. Unfortunately, to stress that what is going on in this book is not an attack on the work of the last three decades in science and technology studies and semiotics, but a change of emphasis; it is necessary to repeat the point that a claim that is not being made here is that longer strings are *always* more likely to succeed than shorter strings. Sometimes longer strings just create more confusion. It is just that *sometimes* longer strings will succeed where shorter ones did not. Noticing and remarking on this indicates the change of emphasis. (Those in the field of science studies will know that change of emphasis was first set out in Collins and Evans 2002 when it was mistakenly interpreted as an attack, what we call "Wave Two.")

An example of condition 4 would be if a very early computer was fed the multiplication instruction along with the calculator program but failed until a new memory unit was physically added to the circuit. In the case of humans—and here we switch examples for the moment—we can imagine a person who is unable to lift a 250-pound dumbbell in spite of a string being transferred which is readable as how to accomplish a "clean and jerk." After six months weight-training, however, the human has developed many new muscles. The string referring to clean and jerk plus the new muscles enables the weight to be lifted.

When fulfilled, the higher-numbered conditions may enable the lower-numbered conditions to be more effective—they create an ability to do more things with strings. The computer is more powerful once it has the extra program and will be able to do kinds of calculations beyond multiplication as a result of only condition 1 communication. Additional memory chips will enable it to do still more types of calculation in response to only condition 1, 2, or 3 communications. The weight lifter will be able to do a number of new physical things triggered by short or long strings as a result of the new muscles.

One of the special features of humans or animals is that the condition 4 example could have been replaced by one that involved the brain rather than the body. Imagine a human who could not get the right answer to the multiplication sum even utilizing the longer string of condition 3 ("move the decimal point one place to the right"), but after six months at a special mathematical behavior training school, learning to do calculations somewhat after the manner of the deaf boy featured in box 1, even the condition 1 string would suffice. The equivalent of the new muscles in the case of the weight lifter is, in the case of multiplication, something new and permanent in the brain, such as new areas of dense synaptic connections, perhaps some new chemical soups, and perhaps some things we do not understand. This is also what happens when animals are trained—for example, the case of a dog learning to respond to the shepherd's whistle. In humans and animals, then, the physical change needed in condition 4 communications can take place in the brain substance as well as the body.

In the case of humans, the learning-calculation-at-school example brushes over an ambiguity. This is that learning at school generally involves communication through the medium of language. In the example it is assumed that the calculative abilities learned are closer to a dog learning to obey a whistle than to full fluency in arithmetic. In other words, the example takes it that the arithmetical abilities learned amount to no more than a kind of wooden ability to apply rules of calculation automatically and unimaginatively. The trouble is that in humans, this nearly always involves a degree of fluency in the language in which the teaching is conducted. For the sake of the exposition this ambiguity will be ignored and fluency will be treated as something separate from merely absorbing instructions.

The fifth condition of communication involves fluency in language, and it applies only to humans (or humanlike things, if there are any).[9] Condition 5 refers only to column 4 of table 1—interpreted communications. It is like condition 4 in that it involves a physical change in the entity that is to be successfully communicated with but the change is of the more fluid and context sensitive kind that is associated with becoming fluent in a language or other social skill. Language needs to be described before we get to condition 5.

The Transformation-Translation Distinction

Language is not the same as strings and the difference between them—their different ontology—is a major principle of the entire argument of this book. It will be referred to as the "transformation-translation distinction."

> **A string** can, in principle, be *transformed,* using lookup tables or their equivalent, into other strings any number of times and transformed back again without loss of *information,* where "information" is a term belonging to the physical world. If there is loss of information in practice it can be more or less remedied, or at least measured, by techniques that are the business of the physical sciences and described by "information theory." **A language** *cannot* be transformed. A language can only be *translated,* and translation always involves the risk of irremediable loss or change of *meaning.* There is no physical mechanism that can be deployed to ensure that losses of meaning are always avoided or remedied. There is no "meaning theory" that, like "information theory," can guarantee complete or almost complete loss-free transmission or even measure the losses.

9. It may be that dolphins and chimps or even some other animals share the interpretative abilities of humans to some small degree. Whether this is true or not matters not in the least for this argument, since the argument covers entities with interpretative abilities; the proper extension of "entities with interpretative abilities" does not have to be settled for the logic to remain coherent. Readers who do not like the existing treatment can, henceforward, read "human" as "human or humanlike."

This distinction is indicated by, among other things, the different ways in which we try to reduce losses in the case of string transformation and language translation. In string transformation we try to create "redundancy" by endeavoring to transmit *the same string* more than once. Even if the strings are not transmitted cleanly but distorted by noise, a mathematical procedure can compare the repeated instances treating similarities between them as "signal" and differences between them as "noise." Attempts to render *meaning* clear, on the other hand, involve repeating the message using *many varied strings* in an attempt to make interpretations cohere; sending the same string over and over again will add nothing to *meaning* transfer, however useful it was in the case of *information* transfer.[10]

The reason for these differences is that strings are essentially meaningless whereas a language is always full of meaning, and meaning is something that relates to the changing ways people live in society. A string is just a physical thing and physical things, the second law of thermodynamics aside, can be changed into one another and back without loss of the information they contain. Even in the absence of the second law of thermodynamics *meaning translation* without loss cannot be guaranteed.

The principle of "no translation without risk of loss" applies just as much to ordinary speech as it does to the common meaning of "translation": loss cannot be avoided with certainty even in the case of conversations *within one natural language*. Imagine I read a paragraph to someone and ask them to rewrite it. Assuming the "someone" was not an expert in memory techniques so that they could not, as it were, "inscribe" the paragraph on their brain, the way they would remember the paragraph would be via its meaning. When they rewrote it, they would almost certainly use at least some different words and this would likely result in a small change of meaning. Repeat the process over and over again and the meaning would change more and more, just as in the game of Chinese Whispers or Telephone. In fact, anyone who transcribes recorded spoken English into written form tries this experiment again and again. Listen to a segment as small as a sentence and type it out *exactly* as you believe to have heard it and there are likely to be a surprising number of subtle changes where the speaker's

10. A point exemplified in television or film comedy scenes when an Englishman abroad tries to get his meaning across to the uncomprehending foreigners by repeating the same message in a louder and louder voice—in this case information transfer is being confused with meaning transfer. More seriously, we also try to reduce meaning loss by "hanging about" for longer with those who want to transfer meaning to us (see chapter 2), or "defining terms" (as exemplified in the section of chapter 4 on the term "cannot."). Perhaps we could begin to measure meaning loss with techniques such as the Imitation Game (Collins and Evans 2007).

phraseology has been unconsciously converted into your phraseology. If you want an exact transcription, you will likely to have to read through the typed version while listening to the words at least a second time to correct the errors.[11] This second time it is not the meaning of the language that is being interpreted but the strings that are being transformed. Perfect transformability can come only with meaninglessness and the second listening is meant to substitute *transformation* for *translation*.

Confounding Strings and Languages

It is hard to keep the distinction between strings and languages apart in practice, because language is normally thought of in terms of conversation or the like and representation as strings is ever present. Language translation or just plain conversation within one natural language consists of three stages.

Stage 1: inscription

In "telling" the attempt is made to represent lived meaning with the inscribed string. For example, in the case of conversation an attempt is made to represent the meaning as a string comprising vibrations in the air.[12] This book does not deal with the teller or transmitter of a string. If it did, the problem would need to be broken into two parts. The first part would be the intention of the transmitter. For example, as explained in chapter 5 of Collins and Evans's *Rethinking Expertise*, under the scientific form of life, an inscriber will try to minimize the scope for interpretative flexibility in the inscription being produced, whereas in some kinds of writing the idea is to leave the final interpretation up to the reader or even to encourage a interpretative tradition. In the "adventurous arts," the producer might intend to create a provocation with many unforeseen interpretations. The second part would be to do with how human transmitters of strings try to design their strings to achieve their ends. This part would have to cover teachers, co-

11. These errors, incidentally, are of a quite different kind to the errors that might be made by a computerized speech transcriber. The automated transcriber would produce something that sounded similar but could be very different in meaning; the human produces something that sounds different but is only subtly different in meaning.

12. The term "inscription" has a central role in the work of Bruno Latour (e.g., Latour and Woolgar 1979). But Latour imputes far too much power to inscriptions, for example, in his idea of an "immutable mobile." An immutable mobile is merely an inscription and is therefore essentially mutable. The term appears to solve a problem but merely disguises the problem of the translation of meaning.

medians, dictators, advertisers, parents, sergeants major, broadcasters, telephone engineers, printers, font designers, book binders, the fashion industry, and so on almost endlessly.

Stage 2: transmission and transformation

This is the move of the string from one person to another that always involves *transformation* of the string from one form to another. A typical series of transformations might involve electrical signals in the brain, to movements of the mouth, to vibrations in the air, to movements of a diaphragm, to changing currents in a coil, to amplified currents in a circuit, to currents in another coil, to movements of another diaphragm, to more vibrations of the air, to movements of the ear drum, to electrical signals in the nerves, to patterns in the brain, and so forth. The media might, of course, include materials such as paper on which marks can be more permanently inscribed, and all manner of machinery and other devices. In ordinary conversation, only the air and elements of the human anatomy might be involved but a series of *transformations* still takes place. This is the domain of ordinary cause and effect.

Stage 3: interpretation

This is the attempt to recreate meaning from the string—to interpret it.

To repeat, because strings are always present when language is being used it is easy to mistake language for strings and therefore easy to mistake the meaningful world of language for the physical world of strings. Furthermore, the work we do with strings mostly involves language and this confounds the two for the opposite kind of reason. Thus it is very easy to think that computers deal with language. But the transformation-translation distinction makes it clear how wrong it is to talk of computer "languages": a computer string can be transformed backward and forward into other strings indefinitely without any loss of information; only "information" in its technical sense, not meaning, is involved. *Transformation of strings* is what happens inside computers. Computers, then, deal with strings not languages! Even though it is said that "this" computer can deal with the language C++ and that "that" computer can deal only with the language BASIC, the term "language" is being used metaphorically; it is all a matter of strings. It is a great shame for philosophers that the integrated silicon chip was ever invented; if all the work that is today done with computers was done with elaborate versions of Charles Babbage's Difference Engine, with its clunking gears and ratchets—and it is only logistics that prevents it being so—it would be much easier to understand that a computer is a physical mechanism. Do a multiplication sum on a calculator and it is all physics from the moment the keys are pressed to the moment the liquid crystal display stops changing! In terms of the metaphysics, it is just like opening a can of beans until you start, metaphorically, to eat them. Likewise, the dance of the bees is not a language; it is a string—it can be transformed back and forth into other strings indefinitely without losses beyond those of the kind dealt with by information theory.

Of course, humans are free to interpret these strings as having meaning and often do. They treat the output of computers as though it was human language and they even speak of "the language of the bees."[13] But this amounts to nothing more significant than the song of the yellowhammer being said to be "a little bit of bread and no cheese."[14] The mistake is to think that such strings have meaning for the entity which is being mechanically affected—the computer or the bee—or that they have "inherent meaning." No string has inherent meaning—it is just a string. Obviously, the string of the yellowhammer does not, in itself, "mean" "a little bit of bread and no cheese."

The confusion is even more seductive because of the way humans solve problems of transformation in practice. Information theory, remember, tells us that even mere physical transformations can involve loss of information unless special precautions are taken. When humans transform strings they often switch from *string transformation* to *language translation* in order to try to rectify *information* loss when there is insufficient redundancy to resolve it. For example, in conversation the inscription onto the air is generally done so badly that it is impossible to recover the strings without loss. An audio-typist is engaged in string transformation but can hardly ever do the job without passing through an interpretation stage—a stage involving language even though only string transformation is desired—imagine an audio-typist working in a language they do not know.[15] This encourages us to confuse string transformation with language translation.[16]

13. Cf. Crist 2004.

14. Is the song of the yellowhammer a sign? It certainly is a string which is sometimes interpreted. To repeat, the advantage of talking of strings not signs is that we do not have to worry about the question.

15. So-called computerized speech transcribers make mistakes because language is not available to them.

16. A striking example of humans' ability to use meaning to recapture strings is provided in box 16, on p. 115.

Enabling Condition 5

It is now possible to get back to the fifth enabling condition of communication. Though there is no way to guarantee that translation in all its senses cannot be carried out without loss, there are ways to minimize the loss. Longer or more carefully structured strings can help minimize loss—this is generally known as "explaining better." More significantly, one can attempt to engender a change in the human on which the strings impact so that existing strings or shorter strings will do.[17] This kind of idea has been discussed as condition 4, but under condition 5 meaning is involved. Languages are always changing as the circumstances and societies in which they are located change. Thus, in condition 5, the change in the human that comprises the acquisition of fluency in a language has to be malleable—always open to adjustment as the society and the language changes. The change cannot be fixed as it is in the case of condition 4.

One example of condition 4 given above was transfer of the ability to calculate to school students using quasi-behaviorist methods. For a student who is not going to become a full-blown mathematician, this change involves a fixed set of abilities analogous to the calculator program that might be fed into a computer; we can imagine it involving a fixed, or nearly fixed, new physical increment to the brain or whatever. On the other hand, if the arithmetic was being learned by someone with ambitions to be a mathematician—that is, someone who continually engages in open-ended discussion about the nature of mathematics—then it would be different. It would be a matter of acquiring fluency in "the language of mathematics."[18]

With language, the context of use can change in important ways many times a day; on a longer time scale, but one that is still short in terms of human life, new terms are always entering and leaving the language. If, in condition 5, the new substance was simply fixed or was designed for fixed responses, the user of the new language would appear very "wooden" and predictable in their use and soon their discourse would become archaic.[19]

17. Another possibility is a change in the transmitting entity but the analysis is not significantly different in principle to a change in the receiving entity so will not be separately discussed.

18. The end point for mathematicians is flexible interpretation whereas the end point for troops undergoing basic training is causal interaction.

19. See the critique of the Chinese Room in Collins 1990, ch. 14, and the discussion in chapter 6 of this book.

Table 2. Five conditions of communication.

Condition 1	The transfer of a string gives rise to a communication.
Condition 2	A transformed string gives rise to a communication though an untransformed string failed.
Condition 3	An enhanced string gives rise to a communication though a shorter string failed.
Condition 4	The transfer of a string plus a significant fixed physical change in the receiving entity gives rise to a communication.
Condition 5	The transfer of a string plus a significant flexible and responsive physical change in the entity gives rise to a communication.

Somehow, the ability that has to be transferred to engender fluent language use has to be flexible—it has to be an ability that can respond to social cues and contexts.[20] To date, the only way we know to engender such a change is through socialization. The conditions of communication are summarized in table 2.

20. The possibility of being able to make quick changes in humans without going through the tedious process of training or exercise is a lasting ambition of humankind. Consider, for example, tribes who believe one can gain one's enemies strength or abilities by eating their brains and body parts or James McConnell's (1962) attempt to transfer behaviors from rat to rat by ingestion of brain puree, holding out the promise that one day we will be able to transfer the ability to speak languages (or do arithmetic) by taking a pill. Mostly, however, these ambitions are more a matter of science fiction. Dr. Bruce Banner found he could occasionally turn into the Incredible Hulk as a result of exposure to radiation; H. G. Wells explored the theme in respect to animals in his *Island of Dr. Moreau.* The possibility of instant transfer of language fluency is provided by the Babel fish invented by Douglas Adams for his *Hitchhiker's Guide to the Galaxy* (1979). Once inserted into the ear, the Babel fish enabled anyone to speak any language.

TWO

Digital Strings, Analogue Strings, Affordance, and Causes

In this chapter the nature of strings is examined in more depth. This chapter can be skipped or skimmed by those who do not feel the need to go into a great deal of technical detail before getting on to the classification of tacit knowledge. The analysis starts with a discussion of the difference between digital and analogue strings. The transformation-translation distinction is shown to apply to both. It is then shown that the distinction even applies to artifacts, which can be seen as strings that represent themselves. It is argued that string transformation and ordinary cause and effect merge into each other.

Digital and Analogue

Strings come in two types, digital and analogue.[1] With analogue strings—for example, pictographs or paintings—the substance of the string seems analogous in some way to the interpretation it may be given. These strings

1. The key distinction between the digital and the analogue is that the former consists of a set of discrete entities, whereas the latter is continuous. Thus my pocket calculator is a digital device because it does its calculations and representation of numbers by manipulating only two types of electrical states conventionally talked about as 0s and 1s; a number has to be represented by some arrangement of a finite set of these discrete entities. My slide rule is an analogue device because numbers are represented as positions along a continuous scale which, in theory, can have infinitely small divisions. For further discussion of the meaning of the digital, see Haugeland 1986 and Collins 1990. In an earlier work (Collins 1990), I claim that the impression of a symbol onto coinage along with the use of base metal turned coins from analogue devices into digital devices. Value in the former was proportional to the amount of gold and coins could be clipped; in the latter the value was fixed by the symbol irrespective of what damage was done to the coins. Note, however, that strings are just strings—there is no difference between digital and analogue until we start interpreting—the symbol on the coin has to be recognized, "0" and "1" have to be seen as something different from the molecules that make

seem to have intrinsic meaning but it will be argued that, in spite of immediate appearances, they do not. This follows from the application of the transformation-translation distinction to analogue strings. Analogue strings, however, do have internal form and structure based on the nature of the materials from which they are made. Digital strings have less in the way of any substance that appears to make a natural correspondence with meaningful things. The form and substance of a digital string seems, and is, more arbitrary than that of an analogue string. Even digital strings lose their arbitrariness, however, when they are subject to regular and stable interpretations in terms of meaning (which is when they tend to be called "signs" or whatever).

To say that strings are not arbitrary, however, is not to say that strings contain meaning in themselves. A string is never meaningful, but *some* strings are more easily available for interpretation in a meaningful way than others. The nonarbitrariness is contained in the word "some." If strings were always arbitrary there would be no subset that was more easily interpretable than any other subset.

Affordance

We say that digital strings are meaningless and (initially) arbitrary because there is nothing inherently "A"-like about the letter A, nor does the A in the word "cat" contribute anything to the meaning of the word "cat". This arbitrariness is still more obvious in the case of binary strings—which consist of only two types of element, such as opposite magnetic or electrical states, or 0s and 1s. Since any string can be transformed without loss of information into any other, including 0s and 1s, the inherent arbitrariness in any string must be equivalent to the arbitrariness in binary strings!

Though strings are without meaning, we often talk as though they were meaningful. We say things such as "I have a French-language cookbook that explains how to make coq au vin." But a book is a physical thing, not a meaningful thing, so since in our terminology language implies meaning, there is no such thing as a "French-*language* book." It follows that a book cannot do any explaining since explaining implies meaning (see the quote from *Phaedrus* in chapter 1). We also say things like "this is a photograph of Ludwig Wittgenstein." But, again, a photograph is just ink marks on paper—a string—and, in itself, is not "of" anything or anybody. The right way to describe the book and the photograph is as follows: I have a book that

---

them up, and the wood of my slide rule has to be thought of as continuous rather than made of discrete atoms.

is capable of being interpreted by some French-language speakers as a set of instructions for cooking coq au vin, and I have a sheet of paper with ink marks on it that is capable of being interpreted by some people as a photograph of Ludwig Wittgenstein.

The second two descriptions are each a bit of a mouthful and they miss out on something—that anything is "capable of being interpreted" as anything else; humans are just so brilliant at interpreting "this" as "that" that saying "capable of being interpreted" tells one nothing about what it is that is being interpreted. For example, I might say, "there is a cloud in the sky that looks like Ludwig Wittgenstein, and when I listen to that piece of music I always remember the sequence of notes by thinking of it as the recipe for coq au vin" (or, the song of the yellowhammer sounds like "a little bit of bread and no cheese").

To make progress, instead of saying "capable of being interpreted," I will adopt the term "affords the interpretation," which carries the implication that there is something in the string that makes it easier to interpret one way rather than another.[2] The good thing about the term "affordance" is its looseness: the first definition of "afford" in *Chambers Dictionary* is "to yield, give, or provide." What "afford" does *not* mean is "determine." We can say that something affords something without meaning any more than that it offers a certain possibility. There may be many other things that offer the same or a similar possibility. But it still allows us to say that another set of things do not offer the possibility, or offer it less readily. If I have £5, I can afford the ingredients for a meal; if I have $7.50, I can afford the ingredients for a similar meal; if I have £1, I can also afford the ingredients for a meal but only with a lot more work (for example, a lot more looking around before I purchase anything)—£1has less affordance than £5 when it comes to buying the ingredients for a meal. And whether I have £5, $7.50, or £1, I still have to do the cooking. One must always remember the cooking, which, in the case of strings, is the interpretation.

To put the point in yet another way, you are standing on the edge of a cliff and, seeing your circumstances, I intentionally give you a little push so that you fall and die. In a literal sense, the push did not kill you; the push merely afforded the conversion of your potential energy into kinetic

2. The term was introduced by the psychologist James Gibson but the usage developed here is specific, particularly in the stress that anything can afford anything if enough work is put into it (see, e.g., Gibson 1979). The usage here is, perhaps, nearer to that of Norman (1988) but certainly differs from his treatment in the nondeterministic nature of affordances (see footnote 22 in this chapter).

energy by gravity, and striking the ground with all that energy is what killed you. If you had been standing a little way from edge of the cliff the same push would have been insignificant. Strings are like this. So long as you have already climbed the cliff (the climbing is something like fulfilling enabling condition of communication 4 or 5), and the circumstances are just so (you are on the edge), the push can afford a significant outcome—which in the case of strings is an intended interpretation. Otherwise, there is nothing more intrinsically meaningful in the string than there is in the push.

The terms "afford" and "affordance" are lazy terms. As I will indicate in due course, these terms merely paper over deep cracks in our understanding—or, or at least, my understanding—of why, given the extraordinary interpretative capabilities of humans, anything affords any one interpretation better than any other. How are meanings ever fixed, or even favored? Luckily, for the purposes of this book it is not necessary to do a proper job of repairing the cracks, but it is appropriate to suspect that something hidden and mysterious is going on whenever the terms "afford" and "affordance" make their appearance; one should, I believe, try to associate a sense of discomfort with these terms.

Accepting that there is a problem, here is the reformulated way of speaking in terms of affordance: "to French speakers, this book affords interpretation as a recipe for coq au vin and to people brought up in certain Western literate groups, this photograph affords interpretation as Ludwig Wittgenstein" (whereas most other books or photographs afford them less readily). This is better than saying the book *tells* French speakers how to make coq au vin or that this is a photograph *of* Ludwig Wittgenstein, because these statements are wrong in just the same way that saying a pound of potatoes, some milk, cream, and flavorings is gratin dauphinois. There is no meaning in the book or the photograph any more than there is a gratin dauphinois in the ingredients.

With the bandage of affordance in the conceptual medicine chest, we can return to the difference between digital and analogue strings. If we imagine some moment when written language was still being invented, there was even-handedness about whether to choose the string "cat" or the string "house" to refer to a house; either choice would have served equally well. That is what is meant by saying that the strings are arbitrary. Nowadays, however, the choice is not even-handed because of the way the use of the alphabet has become embedded in social life. Nowadays, if I want to refer to a house, it is better to write "house" than "cat." Nowadays, "house" affords house much better than "cat" affords house, so some of the arbitrariness has been lost. (And this is another oddity of affordance—it can develop and change with use.)

As a quick reminder of the problem of fixed meanings, however, if I were a dictator with similar ambitions to the one who runs the society depicted in George Orwell's *1984*, I might decide that henceforward everyone was going to refer to houses with the letters C-A-T. I'd probably have to change quite a bit more than a couple of words to make this happen, but it could be done with enough effort, and afterward things would run quite smoothly.[3] Thus we can recapture the sense of the arbitrariness of the relationship between the strings and their interpretation. But the fact is that it would take a lot of work to make the change from "house" to "cat," whereas continuing to use "house" to represent a house takes relatively little work. The idea of affordance as it is being used here is a measure of the amount of work that needs to be done to deliver an interpretation: the less work that has to be done, the greater the affordance of the string or object. To put it in quasiphysics language, affordance is inversely related to interpretative work.

Analogue Strings and Affordance

Now let us see how affordance works in the case of analogue strings. It is still harder to make the case that analogue strings start off with no meaning in themselves. Suppose I adopt the pictograph in figure 1 to represent "house" in my language.

It is tempting to say that it is an appropriate pictograph because it looks like a house. But actually, it doesn't look anything like a house—if you doubt that, just take the book into the street and compare the string with a real house. Nevertheless, we are inclined to say that the pictograph is a good one for representing "house"; it has to do with the fact that it "affords" the interpretation "house" more easily than say, the pictograph in figure 2:

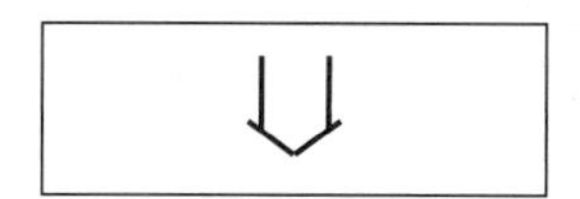

3. Among the extra things that would have to be changed would be all the usages in books since written words are related to the other written words around them; Orwell's Ministry of Truth fulfils this task.

Here the affordance seems to have a little more to do with the string itself than it did in the case of the word "house." It appears, in other words, that even at the imaginary moment when strings were being interpreted for the first time, it would be more "natural" to assign the meaning "house" to the upright pictograph rather than the inverted pictograph. Compare this with the way we assign meaning to "house" and "cat" that has come about only through long usage.

But this is another of those moments when it is good to experience the discomfort that should always come with the idea of affordance. Given that the upright pictograph bears so little resemblance to a house, it is hard to say why the inverted pictograph bears any less resemblance.[4] In other words, the adjectives "upright" and "inverted" should not even apply—neither string looks like a house, so why should we imagine that one of them is upside down? The bandage of affordance covers the open wound well enough, however, for the project of this book to survive.[5]

What makes it worse is that the meaning of many kinds of even analogue string has to be clearly socially established and learned, just as with digital strings and what counts as good affordance changes over time. The string can change and still be recognized even though it would not have been recognized at some earlier time. The string representing "men" as it appears on the door of public toilets—the icon with two "legs"—would have been unrecognizable years ago.[6] Likewise, it took me years—and many detours—to learn that French road signs that point to the left or right (and are angled slightly away from you) are meant to afford the interpretation "go straight on."[7] And the process is nicely illustrated (literally and figuratively), by cartoonists, who slowly train the reading public to recognize

4. This is exactly the same problem of why one thing is said to be an analogue of another thing. To use the idea of analogy you have first to solve "the problem of relevance" (Dreyfus 1996, 173).

5. It could be that the logic of the word "afford" is rather like the logic of the idea of social rule. I cannot say what the rules are for, say, keeping the right distance away from others when I walk along the sidewalk under different circumstances of crowdedness, but I can say when I am not following them and so can every other member of my society. For example, if, on a deserted sidewalk, I brush up against someone of the opposite sex, I could be arrested. If I do the same on a crowded sidewalk, no rule will necessarily have been violated. In the same way, with affordances and pictographs, one cannot lay out the rules for good ones—there are an open ended set of possibilities—but one can recognize bad ones.

6. The example can be found in Chandler 2002. Edgar Whitley kindly sent me a photograph of the sign from a Chinese men's toilet—it shows a smoking tobacco pipe.

7. It seems to me that the English upward pointing arrow has much better affordance but maybe it would be different if I were French. On the other hand, so far as I know, French tourists new to Britain rarely interpret the upward-pointing arrows meaning "go straight on" as

a new Prime Minister in a couple of pen strokes that would have been incomprehensible when he or she was first elected.

It is less obvious still that strings are not inherently meaningful if we think of a photographs and the like. "Surely," we want to say, "this must represent one particular person." But it takes some learning to recognize that photographs are representations of people—after all, they are non-living two-dimensional patterns of ink on paper that are the wrong size, weight, and temperature; unsurprisingly, there are peoples who have not come into much contact with Western society who cannot interpret photographs. Nevertheless, we can still say that a photograph of Ludwig Wittgenstein more easily affords the interpretation "Ludwig Wittgenstein" than the interpretation "Bertrand Russell." Significantly, the fact that few of us have "set eyes on" either Wittgenstein or Russell does not stop us thinking of the resemblance as inherent within the photograph. "That's Wittgenstein," we say—whereas really, at best, we are just "seeing" resemblances between one photograph and another. What we should say is something like, "This string looks similar to me to other strings I have seen that people tell me afford the interpretation 'Ludwig Wittgenstein.'"

The point is made still more clearly by considering the well-known "photograph" shown in figure 3. To those who can see the face of Che Guevara (or Christ) in that image it is as compelling as a photograph; to those who can't, it is merely a set of black and white blobs (or perhaps, as it is said to be, a photograph of a mountain range covered in snow).

Does the image in figure 3 really afford the face it shows to some people? One can see that it could only afford it for those who are used to black and white blobs on paper being interpreted as living humans and, even then, only to those who live in societies where men have beards. To the extent that the image does afford anything, the thing it affords is obviously very much mixed up with the interpretative capabilities and tendencies of the viewer.

In other words, only sometimes does the string in figure 3 feature in condition 1 communication ("look for the face of Che Guevara") and, more often, it can figure in condition 4 and 5 communications. Brief experiments suggest condition 3 rarely applies: if someone can't see the face, it is rarely possible to tell them how to see it however long the explanation goes on.

---

meaning "go upwards," which suggests that I might be right in some frighteningly "scientific" sense. (When newly touring France, I also spent many hours puzzling over maps vainly seeking the town of Poids Lourds—but that's a digital string story.)

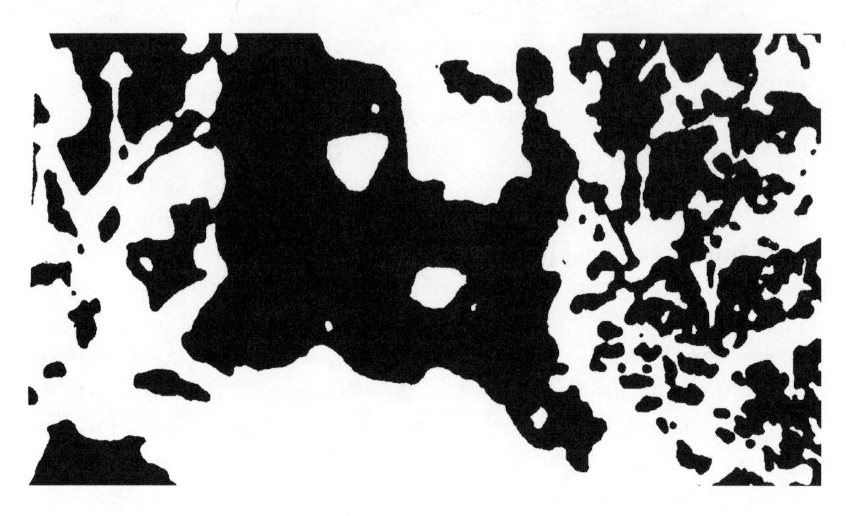

Figure 3. Che Guevara, Christ, blobs, or a mountain range?

Artifacts as Strings

Can the argument about the essential meaninglessness of even analogue strings be carried all the way through to artifacts themselves? The logic suggests that it should. After all, when one looks at an object, one sees just one aspect of it; one does not grasp the whole concept of the thing unless one already knows that concept. When we encounter unfamiliar objects what we see is, as it were, the object as a string that represents itself. Rotate the object into a different orientation and there is a different string! Consider the object shown in figure 4 and imagine looking at it through the eyes of someone born before the 1970s.[8] What do you see—jewelry, perhaps?

Looking at an artifact, then, is somewhat similar to looking at a picture of an artifact—it has to be interpreted before it represents anything. And looking at a picture is like looking at a diagram, which is in turn like reading a description in words. All of these things are strings, in the sense used in this book. Though it is tempting to think that an artifact cannot be merely a string because it really does *contain* meaning, the temptation is to be avoided. Think of the integrated circuit falling into the hands of a computer engineer before the invention of the semiconductor—it would contain nothing usable in terms of information about how it works. Or think

8. Figure 4 is not, of course, an object: it is a string that you are being asked to interpret as an object. You are being asked to carry out a kind of thought experiment.

Figure 4. Jewelry or an integrated circuit?

of a slide rule in the hands of a tribe without arithmetic—it contains no more knowledge than the smile of the *Mona Lisa* without the onlooker. And think of the process known as "reverse engineering." Reverse engineering is taking an artifact apart to see how it works. But reverse engineering cannot be done unless the engineers are pretty near to the point of being able to "forward engineer" the same artifact—otherwise one might as well expect the cave men to reverse engineer the mysterious black obelisk with which Stanley Kubrick opens his film *2001: A Space Odyssey*.

In sum, artifacts seem somewhat more meaningful than analogue strings, and analogue strings seem to be a bit less arbitrary than digital strings, but in logic they are all the same—they are strings without meaning in themselves. On the other hand, the affordance of artifacts and analogue strings is more intricately related to their internal structure and physical materiality than is the case with digital strings.

Transformation of Analogue Strings

Given that we seem to have progressed without discontinuity from strings to analogue strings to artifacts, the transformation-translation distinction should apply to all of them. In the case of all strings—digital, analogue, or artifacts themselves—we should be able to create unambiguous "lookup tables" that can *transform* one set of strings into another without loss. Table 3 is a lookup table where the correspondences between some numbers and letters and their binary code equivalents are shown.[9]

Lookup tables stabilize the correspondence between different strings. Using a lookup table, we can replace a lost pawn in a chess set with a matchstick or bottle top and replace either of these in turn with a new pawn, should we find one. The lookup table tells us that the matchstick, or bottle top, must be replaced with a pawn rather than, say, a rook or a queen. In the same way, we can replace letters of the English alphabet with semaphore or Morse code or combinations of binary symbols. Or we can replace letters of the English alphabet with letters of the Greek alphabet, or any other alphabet, or different kinds of bottle top, though no one will be able to make any sense of these strings until they are transformed back to the original string system by use of the lookup table(s). (There won't always be one-to-one correspondence between number of elements on the two sides of any one line of the lookup table, but combinations of elements will do the trick: binary strings, as in table 3, which have only two elements, are an extreme case, since only two elements suffice to represent any other string).

When we get to analogue strings, things are less obvious. The reason that it is easy to return to the original without loss after any number of transformations in the case of digital strings is that we don't care if the strings we end up with are drawn a little differently to those we began with because these differences have no significance. As far as table 3 is concerned, **A** is the same as **A**, **01000001** is the same as **01000001**, and one bottle top is the same as another bottle top is the same as a matchstick. Indeed, insofar as we treat the strings as digital, such differences will be unnoticed.[10] But if

9. To be exact, if a lookup table is to work for two-way transformations, it must have one entry (which might be unique combination of elements) on the right hand side for every entry on the left hand side. It must be a "one-to-one" lookup table.

10. Actually, if you can see the point of stressing that "nothing has changed" in this example, then you must be treating the **A** and the **A** as something other than digital—you are seeing the difference. Collins and Kusch (1998) describe the way that printed type is not only a digital string but also affords certain other kinds interpretation as a result of the way it is

Table 3. A one-to-one lookup table.

1	01
2	10
3	11
A	01000001
B	01000010
C	01000011

our house pictograph changes, in the course of its journey through a set of lookup tables and back again from the left hand icon in figure 5 to the right hand icon something may have changed.

The changed string might be felt to afford the idea of "house" less effectively than it affords the idea of "hut." This is because in the case of analogue strings the actual structure—and sometimes material—of the string is related in some way to the affordance. Note that the fact that the bottle top representing a pawn in a chess game is made of metal does not make it an analogue string, because nothing turns on the qualities or shape of the metal. If it were an analogue string then the affordance would turn on the qualities and/or shape.

This does not mean that analogue strings cannot be transformed backward and forward indefinitely and without loss; it just means that it has to be done with much more care than is the case for digital symbols. For example, the transfer could be done so long as the relative position of every atom in the ink that makes up the "house" pictograph were recorded and noted in the lookup table. Which is to say, in the last resort, in the world of strings, every analogue string can be reproduced, in principle, by a digital string, if enough trouble is taken—and this has to be the case on the principle that the elements of any string can be transformed into the elements of any other. It is just that this degree of trouble can rarely be taken in practice—it is too logistically demanding—and that is why there is a distinction between digital and analogue for all practical purposes. Exactly the same arguments can be used in the case of artifacts: so long as we represent the exact position of every atom in the integrated circuit, it can be transformed into some other kind of string, transformed back again, and will be

printed: the right margin may be ragged or justified, different fonts have different connotations, and so on. These differences are built up in just the same way as other social conventions.

Figure 5. Icons for "house" and "hut"?

same artifact as it was before. (At this point in the discussion we are clearly dealing with principles, not practice.)

Strings, Meanings and Lookup Tables

Because of the transformation-translation distinction, the relationship between strings and meanings must be different than the relationship between strings and strings. Though strings are sometimes used to represent meanings, their relationship to meanings cannot be stabilized with lookup tables in the way that the relationship between one string and another can be stabilized. This, remember, is because meaning is continually changing as it lives its life in society.

What was probably going on when the "house" string was changed into the "hut" string was a *passage through meaning*. The house string, as it left my society, afforded the concept of "house" as being something with a pitched roof. Its meaning was developed, we may imagine, by ostensive definition—holding the string up and pointing to houses—use in conversation, and the general settling down of the string into the network of meanings that make up a society's relationship with its language. We can imagine that the string was then taken to a society where people lived in huts without pitched roofs that they still thought of as their houses. The idea of house in that society is, in a mysterious way, more readily afforded by the "hut" string. So when the attempt is made to transform the "house" string backward and forward through meaning as an intermediate step, a significant change happens. In the last paragraph it was explained that such changes were not inevitable just because the strings were pictographs but the way the pictographs had to be transformed if they were to be preserved without change was by creating a lookup table that stripped them of meaning: the transformations treated the strings as material entities described as positions of ink molecules on paper—as it turns out these tend to be very long strings. There cannot be a lookup table with meanings in one of the columns, because we don't know how to represent meanings outside of the lived context in which they are found. "House" just means something different in the house society and the hut society, and if the strings are transformed through meanings then the strings might well change. Here, an attempt to effect a *transformation* has ended up as a *translation*.

Life is not complete chaos because the relationship between strings and meanings is stabilized for much of the time through the sharing of taken-for-granted ways of going on in social groups. It is through socialization that we learn to attribute roughly the same meanings to similar strings most of the time and that we can anticipate what will be a string with the right kind of affordance when we start the process of "telling" by engaging in inscription.[11] Those "roughly the same" interpretations that give rise to reasonable anticipations extend across areas of social life which have overlaps in their ways of going on or cultures.[12] Mathematical symbols are very stable because the mathematical way of life is almost entirely uniform across societies that are different in many other respects. The same is true in the realm of science in general, which we understand since Kuhn to be the relatively uniform but still open to mistranslation. The use of strings within a social group, in which all know how to use them with roughly the same meanings, is called "using a natural language" (or sharing a scientific paradigm).[13]

Let us try to make the point in yet one more way. This book in itself contains strings, not language; therefore it does not in itself contain knowledge. This book is like the smile of the *Mona Lisa*. Examine the painting very carefully and you will find that the smile consists of nothing but paint. But the paint is not the smile. The smile is some combination of the paint and the person looking. In the same way, the book is not the knowledge; the knowledge is the book and the person reading it—so long as they are the right kind of person.

And again, if the *Mona Lisa* does not contain knowledge or meaning, should this not mean that the *Mona Lisa* can be transformed into some other string and transformed back again without loss? And the answer is "yes" once more. It is not hard to imagine that we could build a machine that, by scanning the painting with different frequencies of electromagnetic radiation and measuring colors and contours from the reflection, it would be possible to record the composition and position of every atom on the surface of the painting on digital tape. We then imagine this tape being transferred to another machine that could use the string inscribed on the

11. Various philosophical traditions reach this conclusion; the one that influences this treatment most directly is the later philosophy of Wittgenstein (1953).

12. For a discussion of how these overlaps work, see Collins, Evans, and Gorman 2007.

13. Kuhn 1962.

tape to produce a painting that, with the naked eye, was indistinguishable from the original *Mona Lisa*—a kind of perfect, infallible, and almost literal "painting by numbers" that could not be achieved by any human.[14] That the machine has painted a *smiling Mona Lisa* remains, of course, a matter of the paint plus the humans who look at it, not the paint alone. There is no set of rules that can substitute for the ability to see the smile—one cannot explain to anyone who cannot see the smile how it is to be seen—there is, as in the case of the interpretation of figure 3, no condition 3 solution.[15] That is why Polanyi was right to say that "all knowledge is either *tacit* or *rooted in tacit knowledge*." Nevertheless, the term "explicit knowledge" still has meaning: it means a string that, when appropriately transformed, affords, say, the *Mona Lisa* for those who know how to interpret it (that is, those who know how to see the smile).

Analogue Strings and Cause and Effect

In logical principle, there is no difference between analogue strings and digital strings. To be pedantic, all computers, even digital computers, depend on analogues—numbers being represented by electrical states or brass cogs. Nevertheless, analogue and digital computers are importantly different in superficial, if not deep, ways. The two features of analogue computers that give rise to the sense in which the term "analogue" is used here are, first, continuity of process with associated "inaccessibility" (explained in the next paragraph) and second, the fact that the affordance of the strings involved is related to the material of the analogue.

The analogue at the heart of the computer represents something. In an analogue computer it does its representing with a continuous medium rather than a discontinuous medium and, as a consequence, the in-between states, between input and output, are not normally accessible. In the normal way we can know no more about those in-between states—the steps

14. We could say that by this means, the secret of the *Mona Lisa* has been rendered into explicable knowledge (which is not to say that the secret of painting a smile has been rendered explicable; it is only the secret of painting the *Mona Lisa*'s particular, already existing, and frozen-in-paint smile.) Incidentally, something that has been interpreted as being close to this experiment was Adam Lowe's technically assisted full-scale copying of the stolen painting *Wedding at Cana*, now found in the Louvre, and the hanging of the resulting image in its original location in Venice (thanks to Edgar Whitley for pointing this out). See Elisabetta Povoledo, "A Painting Comes Home (or at Least a Facsimile)," *New York Times*, Sept. 29, 2007.

15. This is just another instance of Wittgenstein's (1953) point that rules can never contain all the rules for their own application for each new rule requires another to explain it—the rules regress.

of the calculation or whatever—than we know about, say, the in-between states of, say, a rapid chemical reaction. In digital computers it is easier to access the in-between states because the representation of process is carried out in discrete steps each of which can be separately accessed.

We know, of course, that the continuousness and the inaccessibility of the processes of analogue computers and processes are not a matter of ultimate principle. According to the quantum model of the universe, nothing is completely continuous and, in any case, most analogue processes that are used in computing are discontinuous well above the quantum level—they are based on the behavior of things made of discrete atoms or molecules. And that is why, logistical problems aside, analogue computers can always be replaced by digital computers and any process executed by an analogue computer could be redescribed in terms of the discrete states of the materials from which it is made.[16] This fits, as it must if the argument is to cohere, with the principle that any string can be transformed without loss into the elements of any other string, and since a computer is in a large part a string transformer, its analogue strings ought to be transformable into digital strings.[17] String transformations are physical processes and what is being said here is that the "continuous" physical processes comprising string transformation that happen in analogue computers are reducible to physical processes describable as changes in discrete states.

All that said, for practical purposes analogue computers are important. Analogue computers can be very simple. For example, a chart might use diagonal lines to show the bursting strength of glass tubes of various diameters according to wall thickness (see figure 6).[18] By moving vertically from a chosen bursting strength on the horizontal axis until the diagonal line representing the chosen diameter is intersected, then moving horizontally to the vertical axis, the necessary wall thickness can be read off. Within the limits of the thickness of the lines and the subdivisions on the axes, the first and last elements of the process are continuous, not digital—you can start with any pressure and might end with any wall thickness. Diameter is digital, however. There are a limited number of discrete diagonal lines that represent the wall thicknesses available from the manufacturer who supplies

16. I leave aside the difficulty of engineering a machine with enough digital states to represent the atomic structure of materials.

17. All bets are off when it comes to quantum computers. This is a pretty empty statement, however, as I do not understand how quantum computers work—my knowledge of quantum computers tends toward the beer-mat variety (Collins and Evans 2007).

18. See Collins 1990, 112, for the chart.

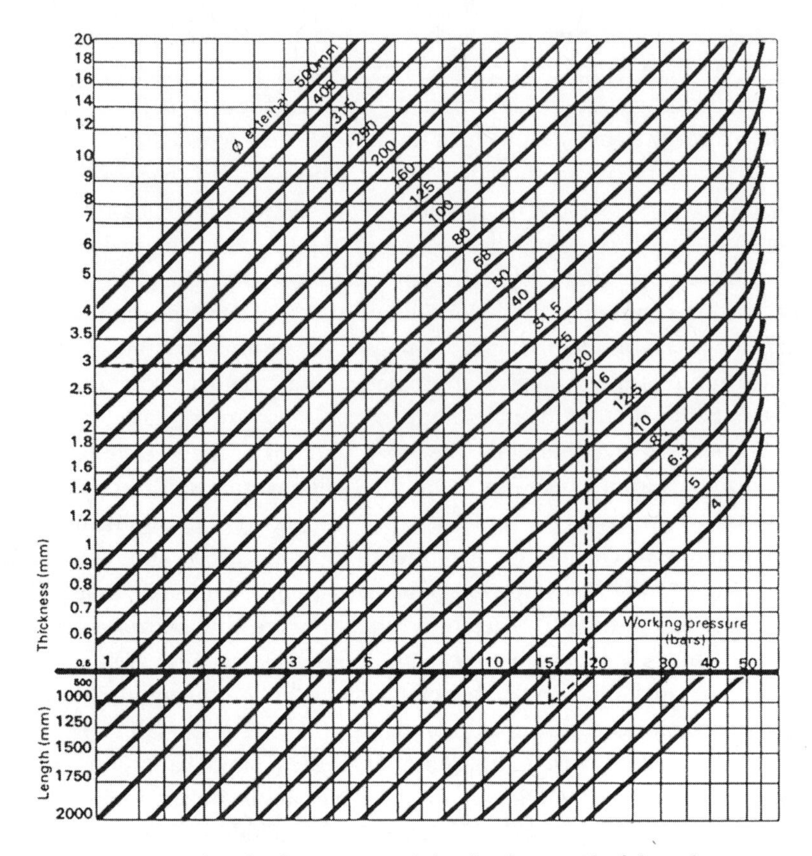

Figure 6. A chart (analogue computer) showing the strength of glass tubes

the chart. No doubt, at order time, one would discover that available wall thickness is also, in practice, discrete.

Notice that figure 6 is both an analogue computer (when used appropriately) and a one-to-one lookup table (and a string).

A slide rule is a slightly less-simple analogue computer. Here the lengths of sliding wooden or plastic sticks are used to represent the logarithms of numbers. Once more, in theory, the numbers that can be entered, manipulated, and read could have indefinitely long decimal tails. In practice, things are limited by the user's ability to read the scales and manipulate the sticks (not to mention discontinuity at the atomic scale). Other examples of analogue computers include Second World War gun directors, which were mechanical devices utilizing the properties of differently shaped pieces of metal as they rolled over each other, and the hydraulic model of the economy built by Bill Phillips in 1949. In the latter, the flow of money was represented by flows of water through tubes and from and into tanks, while choices about spending were represented by opening and closing valves.

Analogue computers of a more playful kind can be constructed with various things such as sticks of spaghetti for sorting numbers, or soap films, which can stretch over a sequence of appropriately placed nails between two boards to show optimum, or near optimum, routes for a traveling salesman to take between towns.[19]

To move to the second difference between digital and analogue computers, though a digital device can be built out of more or less anything, including electrical signatures in silicon or brass cogs, the analogue in an analogue device cannot be swapped at will because other substances may not have the same affordance. The soap film computer works because of the area-minimizing surface tension of soap bubbles—take the soap out of the water and it won't work. The Phillips model of the economy works because water flows downhill and is incompressible—it won't work with gas. The gun directors work because metal is relatively stable and has the right amount of friction—they won't work with chewing gum or Teflon. Rubber bands are not going to be the first choice of materials for making a slide rule. The list could go on and on. This again is equivalent to the difference between digital and analogue strings. The actual structure of an icon has a more intimate relationship to its affordance than the structure of a digital symbol. As before, with enough ingenuity and compensatory mechanisms, analogue computers could be made to work with substances with less affordance but that is a matter of principle not practice.

As intimated, the strings in the analogue devices just described are generally *inaccessible*; the in-between states of the water in the tanks or the positions of the wheels and disks of the gun director are not available for inspection and description (at least, not without the special and logistically demanding effort that would, effectively, turn them back into digital devices). Inaccessibility, as we will see, is one of the sources of the idea of the tacit.

The Metaphysics of Strings, Causes, and Effects

As we have moved from describing digital strings to analogue computers, we should have become more and more convinced that string transformation merges seamlessly into good old-fashioned cause and effect in the

19. Dewdney 1984.

physical world.[20] Babbage's Difference Engine is the most striking bridge between the two ideas. Still more obviously, an analogue computer is just a machine: we pull a lever, the machine clunks and grinds, and something happens at the far end of the mechanical sequence. What happens depends on the analogue. If my slide rule is made out of wood, then pieces of wood have moved; if my slide rule is made out of plastic, then pieces of plastic have moved. If we transform a hurricane into digital strings—that is we build a computerized hurricane simulator with the analogue being electrical states in silicon chips—no one gets wet. Nevertheless, causes and effects that map onto each other are taking place in both instances—both are instances of string transformations, but we just happen to call one a computer because we have control over it, while we call the other a natural event because we don't have control over it and because it is too big to get it inside a building. It also seems harder to model complex calculations with wind and raindrops than the other way round but this again is a matter of logistics not principle, as analogue computers demonstrate.[21]

Computers, of course, merge into the world of machines in general. Imagine I am driving a backhoe (or JCB). I move a little lever with my fingertips and through a series of analogues (for example, movement of cylinders and flows of hydraulic fluids), a much larger arm and bucket moves. Up to a point, the string transformations are invertible (reversible) and beyond that point, when the earth moves, they are not invertible. The backhoe is just an analogue computer being used for something other than computing.

To summarize, string transformations and mechanical causes and effects are, to speak metaphysically, just two aspects of the same thing. This is why we have a strong sense that when we explain some process scientifically we have made it explicit; this is the "explicable" part of the antonym of tacit with its "scientifically explained" connotation. What we have done is take some sequence of causes and effects in the physical world and transformed it into strings consisting of marks on paper. When we cannot re-express

20. Though many-to-one string transformations as well as one-to-one transformations take place in the real world, the potential of indefinite backward and forward transformation is lost. For example, if my computer comprised a giant lookup table containing all multiplication sums the right entry "12" would correspond to the left-hand entries 1×12, 2×6, 3×4, and so on. Such transformations, like many causal processes in the world, are not reversible—or, more correctly, not "invertible."

21. According to *The Hitchhiker's Guide to the Galaxy*, the world is just a big computer run by mice. Incidentally, I am making no claims about real life simulations of hurricanes and so forth being exact. The actual building of a good simulation is not a trivial task (see Collins 2004, ch. 35).

these transformations as marks on paper we say one of two things. If the original transformations are predominantly taking place in the world of things, we say the processes have not yet been scientifically explained. If the original transformations are predominantly taking place inside human beings, we say we have tacit knowledge. In chapter 5, I will suggest that one of the most important classes of tacit knowledge, somatic tacit knowledge, is to be understood as string transformations in humans-as-animals that have not been expressed as marks on paper but could be expressed in principle.

Thus the metaphysical principle of the continuity of strings and mechanical cause and effect explains why we feel that we have expressed some mechanical sequence in symbols we have made it explicit. But the same applies if we take some supposedly "tacit" process in a human (or animal) and mimic it with a mechanism rather than a set of symbols. In other words, if we can make a machine do what a human (or an animal) does we think we understand it—we think it is no longer a matter of tacit knowledge. So if we can make a robot that mimics all the actions of a dog then the mystery of the dog has been dissipated. This, to repeat, is reasonable because, given that mechanical causes and effects are made of the same metaphysical substance as string transformations, then we have explained it. Explicability as modeling causes and effects with strings and explicability through mechanical reproduction are two of the four meanings of explicable which will be set our at the end of chapter 3.

The Components of Affordance

One of the most puzzling things about strings is that even before we can start to talk of their affordance we have to be able to recognize what they are. How is it that **A** and **A** are seen as examples of one string while **B** and **B** are seen as examples of another string? How is it that in whatever old or brand-new ways I scribble the house icon, in whatever color, and on whatever surface, it will be perceived as an instance of the house icon rather than something else? This is a philosophical problem that goes back at least as far as Plato and the cave. How do we come to have stable similarity and difference? How is it that one thing is seen as analogous to something and disanalogous to something else? This is the problem of the meaning of "the same." This problem is not solved here.

Even without this problem, the idea of affordance has plenty of troubles. But there is something that can be said about it. It has three components. The first component is *general physical affordance*—a matter of the physical qualities of strings in general if they are to afford anything at all to humans. An examples is the microdot discussed earlier, which affords nothing to a human until it is transformed by a magnifying glass; a similar example is the whispering sergeant major. More subtle would be the works of Shakespeare transformed into binary code: the string would contain the same *information,* but humans are just not set up physically to deal with binary code—there are just too many elements for us to cope with. Thus, affordance may be lost (or gained) in transformation even though information is not.

The second component of affordance is *particular physical affordance.* This refers to analogue strings and has to do with the way the substance and arrangement of analogue strings seems to relate to the meanings that they afford most easily. This was discussed in relationship to the "house" and "hut" icons. An analogue string can be transformed into another kind of string without losing information but it is likely to lose affordance.

The third component of affordance is *culturally established affordance.* This refers to the way that signs develop meaning with repeated use in society—the cartoonists' drawings, the icons on lavatory doors, and the way one alphabet has affordance in one society and not in another and vice versa.[22]

Computers that Seem to "Read" Strings

General physical affordance is an interesting case, because the work that computers do in rendering strings that do not have affordance for humans due to their physical constitution into string that do have affordance is often misunderstood. The dream of artificial intelligence is that humans and machines—entities that do not share in any culture—will come to be able to communicate in a way that is indistinguishable from the way humans who are part of the same cultural group communicate. One way in which this is supposed to happen is if the strings used for the communication are made long enough so that failures can be remedied by condi-

22. I think that the first two components of affordance correspond to the use found in Norman 1988. Norman deals with the affordance of everyday things, such as door handles. But remember that a door handle wrapped in razor wire still works so long as you wrap it in a cloth before grasping it—this is the equivalent of the extra work of interpretation which is needed when the affordance is low. For the sociologist of scientific knowledge, studying the formation of stable interpretative practices, or the construction of affordances, the indefinite interpretative flexibility is the crucial thing, not the affordance. This feature of affordance relates to component three, which has also been discussed under the heading of sociology of technology—for instance, the way the bicycle comes to have different meanings for different groups over and above its materiality (see, e.g., Bijker, Hughes, and Pinch 1987).

Figure 7. Bridging cultural gaps with long strings "read" by computers

tion 3. The dream is given some substance, because machines are much less restricted than the human organism in respect of the length of the strings they can use. Human brains are slow and human lifetimes are limited, so the subset of strings that have physical affordance for humans has relatively few members each of a restricted length. Machines can deal with much longer and more detailed strings much faster so there are certainly some circumstances where string ramification could accomplish more in the case of machines than it could in the case of humans. A very, very long string that might bridge very the distant interpretative worlds of two groups of humans (if this was all there was to cultural gaps) might be unusable by those humans simply because the reading and interpreting would take too long. A machine, on the other hand, may be able to cope with such a long string.[23]

Computers, however, can transform very long and complex strings that humans cannot handle into strings that humans can handle. They can make transformations that increase the affordance of strings for humans. Figure 7 expresses the point as a cartoon.

In figure 7, messages pass from left to right while cultural distance

23. The conflict between those who accept computer proofs in mathematics and those who think any true proof has to be grasped by the human mind beautifully illustrates the difference, an example that is relevant in several places in the text (Mackenzie 2001).

between people is shown by physical distance on the page. The first line shows two culturally close people communicating with a minimal string—condition 1 transfer, as in the pub joke. The second line shows two more distant people communication equally well with longer strings. This is exemplified by the locals spelling the whole joke out for the visitor—condition 3.

The third line represents a case where the cultural distance is too great to be bridged by a still longer string because the first human can't transmit it and the second cannot process it; it is just too long and complicated. The fourth line shows how the intervention of the computer solves the problem. The string in question might be the works of Shakespeare expressed in binary code, which in turn is inscribed as optical signatures on the disk or it might be the operating systems for the computer. In the first case, the computer can transform the optical signatures back into the Roman alphabet so that it affords Shakespeare to the reader. In the second case the operating system is likely to be so complicated that a whole team of string writers had to get together to put it together and use various transformative and causal mechanisms to inscribe it on the disk. The person on the left could give the CD to the person on the right. But the person on the right can do nothing with it—what the person on the right sees is a shiny disk, just like hundreds of other shiny disks. The string affords nothing to the person on the right except "CD." But slot the disk into a computer and the computer transforms the string into something that does afford something to the person on the right—readable words and pictures on the screen.

To repeat, nothing mysterious happens in line 4 of the figure. The only thing going on is *transformation* of one string into another. The computer does not really "read" anything except in a metaphorical sense; what it does is string transformation. The metaphysics of the situation is exactly the same as if the person on the left had given a plain-language message on a microdot to the person on the right and the computer was a magnifying glass—as in the inset. All that is going on in this example, then, is that the computer is rendering a service that can bring about the second enabling condition of communication.

No *translation* has been accomplished by the machines in these instances. The interpretation of strings does not begin to happen in the examples above until the transformations of one string into another are completed. Translation is often used to describe processes like those above but, though seductive, this usage is misleading; everything that is going on is transformation.

Mimeomorphic and Polimorphic Actions

It is the case that certain of the things that humans do can be mimicked by machines. Does this mean that humans are metaphysically continuous with the world of mechanical causes and effects, string transformation, and so forth? The transformation-translation distinction would suggest not, since humans alone use language and language is not strings. The circle is squared by noting that, first, humans sometimes act mechanically—the example given above is the reflex response to the sergeant major's shout of "'Shun!" In this kind of circumstance humans are easily seen as continuous with the world of mechanisms. The second feature of humans is that they sometimes choose to act as though they were continuous with the world of mechanisms even if the actions are in fact intentional—that is, mediated through the world of meaning.

Actions that are intentional but nevertheless mimic the world of mechanical cause and effect are called mimeomorphic actions. There are many examples, but a good case is the salute. A competent salute is executed as near as possible every time with the same behavior; we could, if we were crazy, build a saluting machine which would salute even better than a human.[24] Actions where the associated behaviors are responsive to context and meaning are called polimorphic actions. An example that is close to the salute is the greeting. If, in the normal way of life, I always greet someone with exactly the same string whatever the context, it would soon cease to count as a greeting and become a joke, an insult, or an indication of madness on my part. As far as an outside observer is concerned, and as far as the possibility of mimicking them in machines is concerned, mimeomorphic actions are not distinguishable from string transformation or mechanical cause and effect, whereas polimorphic actions are distinguishable.

The distinction between these two kinds of human actions has been worked out in detail elsewhere, but because the idea will recur in this book here is another brief summary.[25] Mimeomorphic actions are actions that can be reproduced (or mimicked) merely by observing and repeating the

24. In Joseph Heller's *Catch 22*, Lieutenant Scheisskopf is so concerned with keeping his men in line so as to win a marching parade that he considers "nailing the 12 men in each rank to a long . . . beam . . . to keep them in line" (p. 83). This is the conceptual equivalent of a saluting machine.

25. Machines mimic actions rather than reproduce them, because they do not have intentions. For a complete explanation of the notion of mimeomorphic and polimorphic actions, and many more examples, see Collins and Kusch 1998.

externally visible behaviors associated with an action, even if that action is not understood. In general, in the case of mimeomorphic actions, the same action is executed with the same behaviors (given the zone of tolerance around "the same"). In the case of polimorphic actions, on the other hand, the same action may be executed with many different behaviors, according to social context and, again, according to social context the same behavior may represent different actions. Therefore, polimorphic actions cannot be mimicked unless the point of the action and the social context of the action is understood. So far as we know, the only way to learn to understand the point of such actions and the meaning of the context is through socialization—immersion within the relevant society.

An example of mimeomorphic action that fits with the example of the sergeant major and the troops is the response of novice troops who have not yet developed the reflex reaction to "'Shun!" Novice troops who are appropriately scared of their sergeant major will self-consciously try to act as though they had been affected by a mechanical cause when they first hear the command; they will use their interpretative abilities to understand the order as "come to attention" and then carry out the order as mechanically as possible. They will try to act out of an intentional act as though it was a matter of mechanical cause and effect.

Reprise

The aim of this chapter has been to understand the relationship of digital to analogue strings. The attempt has been made to preserve the transformation-translation distinction in the case of analogue strings and artifacts—that is to show that even these contain no inherent meaning though they do have differential affordance. An attempt has also been made to show the continuity of string transformation and physical cause and effect.

THREE

Explicable Knowledge

Drawing on the analysis of the last two chapters, it is now possible to describe some of the ways the term "explicit" and "explicable" are used in the world as we know it. "Explicit" is something to do with something being conveyed as a result of strings impacting with things. Much of the work of this chapter is to be done through the attempt to answer the list of questions posed in the introduction by using the analysis developed above. As the answers emerge, we will uncover four meanings of "explicable." First, though, the many different ways humans use strings are set out.

Use of Strings by Humans

Strings that Work with No Tacit Contribution from Sender and Receiver

Suppose I'm watching a live soccer game and you are obstructing my view of the match and I want you to move a little way to the right. I can simply push you to the right using as much force as necessary to move your bulk. Is the push the impact of a string? One has a choice about what to call it. Earlier it was argued that the mechanical efficacy of strings merges with ordinary old-fashioned cause and effect in the physical world, so one could call it a string impact if one wanted or one could call it a mechanical effect. One could not call it part of "telling," because it is not interpreted. A gentler push—a nudge—might be part of a telling; one pushes gently, without enough force to move the bulk of the other person but merely indicating that a move is desired. This gentle push is a string which can be interpreted as meaning "please move aside" and can therefore be seen as the middle element (stage 2) of telling. Of course, the person pushed might interpret the hard push as part of a language signifying aggression, but that is another story. The hard push is best seen as part of a continuum of mechanisms that result in movement which include crude force at the one end and more subtle effects such as those associated with halitosis or pheromones. It is the hard physical, obviously mechanical, push that is the main point of this paragraph.

Just a step up from this is habituation or training of the kind used in the case of animals or soldiers in boot camp. Imagine newly captured slaves being "trained" to row a galley. The galley master and the slaves share no language except the "language of the whip" (not really a language at all) and access to food and water. The slaves are trained though the mechanical effect of punishment and reward and, as far as the slaves are concerned, meaning and interpretation is not involved. A little later in this chapter this kind of training will be discussed again in the context of neural nets.

Strings that Require that Sender and Receiver Share a Language

The next level of communication is illustrated by the following description of how a hologram works:

> A hologram is like a 3 dimensional photograph—one you can look right into. In an ordinary snapshot, the picture you see is of an object viewed from one position by a camera in normal light. The difference with a hologram is that the object has been photographed in laser light, split to go all around the object. The result—a truly 3 dimensional picture!

This description comes from the back of a beer mat found in pubs around 1985 that was used to advertise the drink Babycham. It is unlikely that anyone would learn this as a chant.

What would the beer mat "mean" to a typical reader? This question can be converted into something more tractable by taking up Wittgenstein's advice to ask for uses, not for meanings. What could the typical reader do as a result of reading the beer mat that they could not do before they had read it?

To see how this approach works, it can be applied to the examples of "communication" dealt with above. The person who was forcefully pushed aside so as to provide me with a better view will be able to do nothing new as a result of the experience. The regime of punishment and reward offered to the slaves will shape their actions for rowing a galley, though the slaves themselves need have no idea that they are doing something called rowing nor what its purpose is; the slaves' new abilities are like the new "abilities" of a heavy stone that is taken up and used to prop open a door—the stone has "learned" to prop the door; the slaves can do nothing with the knowledge except row what they may not even know is a boat. The deaf boy who learned to copy the English sentences, and me, when I learned to chant the multiplication table, were in roughly the same position as the slave or the stone, the difference being that our new abilities were expressed in written or verbal form rather than physical manipulations—though they were really no more than physical manipulations which happened to produce marks or sounds. Note that I was no more able to do arithmetic in virtue of having learned to chant the tables than I would have been by having them inscribed on my forehead—there is a lot more to arithmetic than knowing the tables. The fact that in none of these cases has much in the way of new abilities been acquired fits with the sense that not much meaning had been transferred in any of these instances.

Turning back to the beer mat, what would the typical reader be able to do as a result of reading it that they could not do before? They would *not* be able to make a hologram, repair a hologram, invent new techniques of holography, and so forth. They might, however, be able to answer questions in a general knowledge quiz along the lines, "Are lasers used in holography?" To be used even in this way, however, requires that the relationship between sender and receiver be a little more close than it was in the case of the deaf boy or the person who can chant the multiplication tables, who (as in the case of the multiplication table slave who follows me around) might have no idea what the tables mean because they do not even have to speak the natural language which the number string affords. To be used even to answer questions in general knowledge quizzes, the natural language, which includes the ideas of "photography" and "laser," must be shared by sender and receiver and those technical ideas must have some meaning to the beer mat reader. For a person who was not an English speaker, reading and remembering what was on the beer mat would not enable them to answer general knowledge questions. Here, then, is the first category where the sender and receiver need to share what is conventionally called "tacit knowledge"; in the cases discussed before we reached beer-mat knowledge, an animal with suitable vocal or physical abilities—a parrot, or an ape, or a human-as-animal, in the case of the galley slave—could do the job: the communication was useful irrespective of whether the receiver had any tacit knowledge. In the beer mat case, the communication is useful for answering questions in general knowledge quizzes only so long as the receiver does already have the knowledge needed for the inscribed strings to have the right affordances and for the questions which might refer to the beer mat to have the right affordances.[1] Chinese rooms and the like aside (these will be dealt with later), these abilities are normally transferred through socialization—the only known means of condition 5 communication. And since they cannot benefit from condition 5 communication, a parrot, or an ape, or a human-as-animal could not do the job.

More gradations of this kind now follow. A technical paper on holography published in the *Review of Scientific Instruments* might be thought to be different than the beer mat in the same way that an entire funny story is different than the number that represents it. But it is only different in this way so long as the reader has the technical language that would allow the potential extra affordance in the extra string(s) to do its job. Given the technical language, the paper might allow the reader to " to make a hologram, repair a hologram, [and] invent new techniques of holography." To use the paper in this way requires that the reader has "specialist tacit knowledge."[2] The relationship which we saw illustrated in the pub-joke joke—the closer the cultural relationship between sender and receiver, the shorter the string can be and still do its job and *vice versa*—applies only under a *ceteris paribus* clause: both parties must share the language, which allows a longer string to afford a meaning. There will be occasions, such as when the typical pubgoer (who possesses "ubiquitous expertise" but none of the specialist tacit knowledge of physics) is presented with a technical paper on holography instead of a beer mat. Such a typical pubgoer will not be able to benefit from the longer string.

Some Varied Uses for Strings

"Asking for the use" invites another dimension of analysis which is orthogonal to the relationship of length of string and technical competence. Consider the two descriptions of goal-kicking in rugby found in box 2. The two descriptions have very different kinds of use.

The first set of instructions might well enable some human somewhere who could not initially kick a goal at rugby to be able to do it. This will depend, as always, and as Webb intimates in the last sentence, on whether that person had the necessary foundation of tacit knowledge and, of course, the right kind of legs, muscles, and so forth, or whether they are in need of

1. This is "ubiquitous tacit knowledge" in the language of the Periodic Table of Expertises (Collins and Evans 2007).

2. Seen Collins and Evans 2007, ch. 1. For a paper in *Review of Scientific Instruments* failing to work in this way, see Collins 1985.

Box 2. Explicit instructions for kicking a goal in rugby football?

"My ritual, my routine? . . . Always adjust the ball so the nozzle is slightly to the right, just so, then angle the upright torpedo just slightly towards the posts for more forward impetus. Eye alignment, Inner calm. Consider the wind. Stand. Left foot alongside ball, right instep right behind. Visualize the kick sailing over. Eye and foot aligned. Wipe hands. Four precise steps back. Stop. Check. Visualize. Then two-step chasse to left if it's a Mitre ball that we use at Bristol, one-and-a-half it it's a Gilbert at Twick[enham]. You must time them better though they go further. Visualize how it will feel on your foot . . . . If you tried to write down on paper exactly what you do to kick a ball between two posts with absolute certainty, it would be impossible, you'd still be at it in a million years—but once you've done it just once, your body and mind has the exact formula stored and ready to be repeated." (Jonathan Webb, then fullback for the England rugby team, quoted in the *Guardian*, March 12, 1988)

"The rugby full-back has to know his co-ordinate geometry. His approximate locus for the wider conversion must be a rectangular hyperbola on which the kicker must place his ellipsoid of revolution before sending it on its parabolic way." (D. Crothers, "How to Pot Black with Some Angular Momentum and a Touch of Gravity," *Guardian*, August 27, 1987)

a condition 5 transfer in order to use the string. Webb's phrase "your body and mind has the exact formula stored," is exactly to the point, and we will return to it in chapter 5. For the moment, however, to the extent that it could help someone kick a ball—to the extent that it could work as a condition 1 transfer—it is likely to be referred to as explicit knowledge.

The second kind of instruction, on the other hand, does not seem usable by any human goal kicker. Instructions such as the second string will also be discussed at greater length in chapter 5, but for now we need only ask whether they are a candidate for the conventional "explicit knowledge" category.

It would certainly seem odd if we were to say that scientific knowledge is not explicit knowledge; science seems the quintessential way of uncovering hidden mechanisms and making them explicit. One way to resolve the problem is to agree that this is a string that, though it does not afford the same use as the previous extract—rugby ball kicking by humans—does afford an understanding of the ballistics of rugby balls. Thus, the knowledge that the string affords could be used for passing an examination in ballistics. To anticipate chapter 6, this is explicit knowledge that will help those who take part in the discourse of ballistics as applied to sport and might be a useful scaffolding for acquiring "interactional expertise" in that specialism.[3]

The second way of looking at the matter is that this scientific string might well be usable by the designers of a rugby ball kicking machine. It could be a high-level string that could be broken down into a ball-kicking program. It just happens that humans are not the kind of mechanisms that can make use of this type of string in the way that some machines might—a possibility that will be discussed at length in chapter 5.

We can say one thing for certain: one of the meanings of "explicable" (the fourth definition in table 4, below) is "can be explained" and explaining is what the second passage does. That the other meaning of "explicable" is the opposite of "cannot be made explicit" just illustrates why the question about the rugby ball is so confusing when we start from the existing discourse: the second rugby ball kicking instruction certainly fits one meaning of explicable and almost certainly does not fit the other.

Second-Order Rules, Coaching Rules, Hints, and Tips

Consider the process of teaching someone to ride a bike. The following advice will be helpful (assuming a shared language):

- It should take between half an hour and half a day to learn to balance reasonably well.
- It's harder to balance when you are going very slowly.
- Try to look well ahead when you ride, not down at the ground immediately in front of you.

The first of these is what has been called a "second-order rule." A second-order rule tells one something specific about how hard the skill is—a very useful piece of information if the skill is to be acquired. There are two examples in box 3.

A second-order rule can be something that can be communicated with a minimal string—("about half a day"; "about six iterations"). The strings are not much more ramified than the numbers that indicate jokes in the

3. The idea of interactional expertise (e.g., Collins and Evans 2007) will be discussed at greater length in chapter 6.

Box 3. Second-order rules of laser building and ferret spaying.

"We regularly tried to build helium-neon lasers in the lab for staff projects. And, if you didn't know that this laser could lase, you would never believe it; it requires such patience to get it started. It makes you wonder how he [the inventor] . . . ever got it to lase because it requires so much patience to line up. Once you know it will go you can do it." (Scientists quoted in Collins, "Tacit Knowledge, Trust, and the Q of Sapphire," 74)

"[The veterinary surgeon has the] experience of finding the uteruses of other ferrets. He knows from this experience that he should find the uterus almost immediately. In short the surgeon brings to the operation some idea of how long it will take and how many times he should repeat the mapping procedure before giving up the search. We are not suggesting that these measures are exact, but measures do not have to be exact to be useful; the crucial thing is to know for sure when you are well outside the acceptable window of a search . . . . [W]e can say that in this case if the surgeon spends over an hour looking for the uterus he knows he has spent too much time. If, on the other hand, he abandons the search after ten seconds he knows he has not spent long enough. He also knows that one failed search for the uterus will not be enough, but fifty searches will be too many. From the video-recording of the ferret surgery it seems that in this case six searches was considered adequate. [In this case the surgeon decided after about six searches that the pet ferret must already have been spayed and that was why the uterus could not be found.]" (T. Pinch, H. M. Collins, and L. Carbone, "Inside Knowledge" [1996], 175–76)

pub. Yet these strings have a use: they help with the learning or execution of a tacit knowledge laden skill. They are useless, of course, to anyone who is unable to master the skill in the absence of condition 4 transfers—for example, someone who cannot cycle because they are wheelchair-bound or who cannot execute the instruction that there should be six iterations when looking for a lost uterus because they have never done a surgical operation.

The second two bulleted examples are of a type called "coaching rules," or they could be called "hints" or, to use Wittgenstein's term, "tips." Like the second-order rule for bicycle riding, such rules are organism specific. They are not the rules for, say, balancing on a bike such as might be programmed into a bike-riding computer but rules to help humans learn to ride on bikes in the way that humans do it. A coaching rule in golf is to hum the "Blue Danube" so as to get the right rhythm for the swing. Nothing could be more organism specific than this. If, as in the case of second-order rules, they could be used in condition 1 communication (suddenly making someone who could not accomplish something accomplish it), then it is not unreasonable to refer to them as explicit knowledge in ordinary discourse. They can, after all, be written down, carried around by intermediaries, and used to improve your riding, your game, or whatever.

Hints, too, could be called explicit knowledge, though this is not the sense of explicit that connotes scientifically explainable, nor is it explicit in the sense of being a transferable version of the abilities to balance a bike. And, of course, hints are not all there is to learning a new skill. The crucial thing is to not allow the language of explicit knowledge to be "taken on holiday": it must not be used in technical settings for which it is not exact enough, or used to draw inferences that it cannot support. The language of strings and enabling conditions of communication is much more exact, much less confounded, and much better for these purposes.

Strings that Are Not Part of Languages because They Are Entirely Contextual

Consider a long-married husband and wife about to leave a dinner party: a glance, a grunt, or a look can carry a wealth of meaning. That there can be a meaning in such a minimal string is a result of years of mutual interaction, but it does not seem right to say that is a merely a mechanical effect because it is not a matter of training in grunts and glances—there is no fixed, reflex-like reaction to them as there is to the sergeant major's shout of "'Shun!" They seem more like terms in a language than simple strings because they are very likely to be misinterpreted if transformed, as their interpretation is such a context-bound matter. Each nod and look takes its meaning only from the particular circumstances in which it is used. Indeed, the husband and wife can communicate with "looks" and grunts that neither of them has seen before and neither of them will use again, and that is why there can be no dictionary of looks and grunts. The looks and grunts of one couple cannot be mastered by the new wife of a divorcé; we cannot imagine a book of looks—a lookup table that converts them into some other known string—that could be handed over to the new partner.

But if the looks and grunts are a language, they are language of an extreme kind—probably best not referred to as a language. There are no publicly available agreements for what they mean; there are only agreements made there and then for the particular context and moment of use. They are, in short, the most private languages that can exist, and they seem too local in meaning to count as language proper.[4] In this sense, the grunts and nods are more like mechanical causes in the way they are used than terms in a language but they are not mechanical causes: they are just an interestingly anomalous case on the boundary of mechanism and interpretation.

Human-to-Human Communication Summarized

In the above section, examples have been provided of the different uses of strings in human-to-human communication. The initial examples involved no tacit knowledge on the part of sender and receiver—these examples were physical pushes, pheromones, halitosis, punishment-and-reward regimes, and the recitation of memorized strings. The exposition then moved on to examples where ubiquitous tacit knowledge—knowledge of a natural language—was required to make it possible for a string to be used; the example was the beer mat. Then specialist tacit knowledge was brought in—necessary to make good use of a technical paper. Then examples were given of different kinds of use—second-order rules and coaching rules—that were explicit knowledge in the sense of being usable by those who had the appropriate physical form but depended on the transfer of tacit knowledge by some prior condition 4 mechanism. Finally, examples of communication were given where nearly everything that was tacit was already in place—the glances and nods between husband and wife—but the instances of communication were so specific to the moment that the term language did not seem to apply but they were still maximally context dependent.

Questions about the Meaning of "Tell" Revisited

In the introduction a list of puzzling questions was posed; we are now ready to answer them. Answering these questions is mostly a matter of clarifying concepts and language rather than providing new information about the

4. The Wittgensteinian private language argument (Wittgenstein 1953) concerns whether an individual can possess a private language; an often-stated conclusion is that this is impossible because a language requires public agreement over judgments of use and this requires at least two persons (for a discussion, see Kusch 2002, especially the concept of "duettism"). But a language owned by only two persons does not seem like much of a language. What one might call the "public language argument" should be about how big a group of people are required to support a language with kind of richness and ambiguity we normally associate with language as opposed to a system of strings.

world. It is mostly a matter of resolving or dissolving puzzlement about the way we use words. The resolutions are accomplished using what has been discussed in the above chapters, and some of the answers are merely repetitions of what has gone before. For this exercise the questions and answer are divided into four groups.

(1) Questions about Encryption and Deciphering

If I encrypt the map reference for a U-boat rendezvous, do I know more than I can tell and is the encrypted message explicit or tacit knowledge? It seems odd to call an encrypted message explicit knowledge. The whole point is that it is concealed. But it is certainly explicable, so in spite of being concealed it does not seem to be tacit: the purpose of writing down a coded message is to afford something to somebody. So it is a string and just like any other string—there are people to whom it does not afford anything. But it will afford its meaning to a subset of those same people if they can decrypt it: decrypting is just a complex and difficult piece of *transformation* that turns on the discovery of the right lookup table.[5] Whether one wants to call an encrypted string "explicit" or not is a matter of preferred English usage—the *Chambers Dictionary* definition of explicit includes "not merely implied but distinctly stated; plain in language; outspoken; clear." It is not explicit in that sense. Furthermore, given that some people want to keep the message secret from some other people, the encoded message could be said to be "relationally tacit" (see chapter 4) in respect of those people who want to read it but can't. Again, there is no real problem here if we stick to the idea of strings and transformations but one can see how the language of tacit and explicit can cause puzzlement.

Before the Rosetta Stone was translated, could its contents be told? And how did I know that it contained knowledge of any kind and wasn't just a pretty pattern? The Rosetta Stone is not dissimilar to the case of the U-boat string. Once more, a (fallible) lookup table has been invented for it as a result of passing the strings through the level of plausible meaning. Before the lookup table was put together it would have been hypothesized that it could afford a

5. It is an interesting case because the process of transformation from encrypted to clear—that is, the process of discovering the lookup table—usually takes a journey through the world of language. The code breakers might start by looking for pairs of strings that occur with the same kind of frequency as certain meaningful words would be expected to occur in the original messages. Effectively they are creating a lookup table between *meaning* and strings. Such a lookup table is bound to be fallible so the claim that translation through meaning without loss is impossible has not been challenged.

Box 4. Is this string reading?

A father-and-son team from Edinburgh think they have found a secret piece of music hidden in carvings at a famous medieval chapel in Midlothian:

> We were convinced from the position at the top of the pillars of the angels and they are all directly under the arches where the cubes occur that there was music there.
>
> We got clues from other books as well. Over the years this became more of an obsession than anything else and we decided we had to find out what was going on.
>
> What we have here is a recorded piece of music, it is almost like a compact disc from the 15th Century.

A reader comments:

> Music, like mathematics, is based on patterns. . . . [T]he harmonious proportions they have found in the architecture of Rosslyn Chapel are no more significant than those found in a '50s high rise. Perhaps the magnificent symmetries of the Forth Bridge will provide the inspiration for their next opus.

(Exchange on the BBC Web site, http://www.bbc.co.uk, April 30, 2007 [Thanks to Neil Stephens])

meaning because of the way the elements were laid out and from previous encounters with such artifacts—that is, we would have guessed that someone, somewhere, sometime, had interpreted the string in a meaningful way. It was this kind of surmise that made the work of symbol transformation—with its many trials and errors—seem worthwhile. But the surmise could have been wrong.

The numerologists who read messages into passages of the Bible or those who can hear satanic messages when certain pop music recordings are played backwards base their claim that difficult to penetrate codes are present in material on the grounds that the meaning they find in their interpretations was intended by someone when they first inscribed the string. But they are probably wrong. The hypothesis that someone or something intended the interpretation that a person places upon a string is perilous, as is shown by the song of the yellowhammer and the case found in box 4.

On the other hand, sometimes what initially appears to be merely a haphazard arrangement of shapes—a string with no ready affordance—can

Box 5. This probably is string reading.

"Humboldt was standing in front a gigantic stone wheel. A whirlwind of lizards, snakes' heads, and human figures broken into geometric fragments. In the center, a face with outstretched tongue and lidless eyes. Slowly the chaos resolved itself; he recognized correspondences, images that enlarged one another, symbols repeated at minutely regulated intervals, and that encoded numbers. It was a calendar."

"Shortly afterward, as Humboldt packed up his instruments, he knew that on the day of the solstice, the sun when seen from the highway rose exactly over the top of the largest pyramid and went down over the top of the second largest. The whole city was a calendar. Who had thought it up? How well had these people known the stars and what had they wanted to convey? He was the first person in more than a thousand years who could read their message. (Daniel Kehlmann, *Measuring the World* [London: Quercus, 2006], 171; 176–77)

appear to yield a compelling interpretation when enough ingenuity is applied. Perhaps what looks like a random pattern only starts to afford the intended meaning when you extract every sixth mark; perhaps the exact spacing of the stones in Neolithic circles really are telling us something. The point is illustrated twice in box 5.

Another example close to my heart is the popular song from 1950s Britain, the first phrase of which goes: "Mairzy doats and dozy doats and liddle lamzy divey/a kiddley divey do, wouldn't you?" As a child, I learned this chant from the radio and sang it with gusto along with my mother and her friends, believing it to have no more intrinsic meaning than the sequence of notes of a songbird. Years later, I learned that this phrase was a string, which, appropriately interpreted, affords information about the eating habits of herbivores (repeat it a few times while thinking about this)!

Another particularly interesting case that exhibits all the dilemmas is that of "dolphin language." Are the sounds made by dolphins just a display of a complex pattern, like the notes uttered by a songbird, or do the sounds afford complex meanings to other dolphins like a language? Only interpretation in terms of our language is going to give us a reasonable indication, and it is going to have to be a convincing interpretation—something more than numerology. Those who devote their lives to the attempt to interpret dolphin language need good reason for sticking with their hypothesis.

(2) Questions about Cultural Distance and Strings

What if a mathematical ignoramus is told some rules for solving differential equations that he or she cannot use?—Is this telling or is it not telling? The rules do not afford equation solving, but perhaps they could be learned to the level of beer-mat knowledge and used to answer quiz questions. Perhaps the ignoramus might be able to memorize them only well enough to recite them and pass them on to others to whom they did afford equation solving. Is it telling? It depends how one wants to use the world "telling," but there is no philosophical difficulty here so long as one speaks of strings, inscriptions, communications, affordances, and the different uses to which a string can be put.

What if I overhear a few remarks exchanged between two people that I can't understand, but, noticing my puzzlement, they explain at length what they were talking about? Is that tacit knowledge being converted to explicit knowledge? And what is meant by "understand" in this context? This question matches the pub joke. A longer string affords something that a shorter string does not because longer strings fulfill condition 3 of communication and can bridge longer cultural distances. Of course, if the cultural distance is too great—for example, the listener has no sense of humor—then the *culturally established affordance* that is required is not going to be established. A change associated with condition 5 is called for (the subject of chapter 6).

What if I give my love a single red rose? Is this telling her something? In our society, passing a single red rose to "my love" (a term that affords the idea of a close relationship) is transmitting a string with a culturally established affordance. The string can be passed by a middle agent (for example, FTD or Interflora). It would seem odd to say to one's love if she asked whether she was loved, "Well I sent you a red rose, didn't I? What could be more explicit than that?" but it is indeed just interpreted communication via a string.

(3) Questions about Computers, Devices, and Animals

If I don't know what is on a computer CD, but I find out when I place it in the drive and the computer fires up, has something tacit been made explicit? Here, nothing tacit has been made explicit; it is just another case of string transformation, in this case a complicated piece of transformation. A string that afforded nothing to me has been transformed into a string—words and images on a screen—that does have substantial affordance, provided my cultural distance from the writer(s) of the string is not too great. Note, however, that affordance is not meaning: if one wants to say (mistakenly) that the string has meaning, it has just the same meaning when inscribed on the disk as when its transformation appears on the screen.

A similar process is unfolding right now as I type these words. The electrical signals squirming about in my computer's circuits, which are part of the process that is producing the symbols on the screen as I type, are inaccessible to me. When the symbols appear on the screen, however, they are in the form of an accessible string. The computer is a "decoding machine" in the narrow sense of a machine for deciphering—something like the machines invented at Bletchley Park in the Second World War to "decode" U-boat messages. In the language of this book, what is going on is "transformation of an inaccessible (to me) string into an accessible string."

Have the programmers told a pocket calculator how to do arithmetic? It is tempting to think and say that a calculator "does arithmetic." On the other hand, the metaphysics of strings and languages outlined here tells us that the calculator is merely transforming strings. Is it, then, that arithmetic is merely the transformation of strings? Part of arithmetic, and for that matter part of the rest of mathematics, is exactly that—the transformation of strings into other strings with better affordance. And it has turned out that part of arithmetic/mathematics is best done by computers: what was once thought to be the epitome of human ability—the rapid transformation of strings, which is known as "mental arithmetic"—has turned out to be a trivial task for machines and as a result, not so clearly signifying superior human qualities. But this is not the whole of arithmetic. Polanyi's claim that "a wholly explicit knowledge is unthinkable" bears on the matter.[6] The claim as stated is evidently wrong because explicit knowledge without tacit knowledge—string transformation—is exactly what we are thinking about right now. If we replace Polanyi's claim with "strings must be interpreted before they are meaningful" and we forget about the word "unthinkable," a level of mystery disappears: we have strings and we have interpreted strings. This idea has a role to play when arithmetic is used in the world. Calculators do not do any calculations except when they are used by humans to calculate. The user has to decide what to calculate and how to use the answer. This requires a background of tacit knowledge, as is illustrated in box 6.

A good way to think about such things is in terms of "action trees." Even particles of human activity that we might be inclined to say are entirely explicit knowledge-based—such as chanting the multiplication table—take

6. Polanyi 1966, 195.

Box 6. Using a calculator depends on tacit knowledge.

The sum "7, divided by 11, multiplied by 11" when punched into many pocket calculators will give an answer with a long decimal tail, whereas it is obvious to anyone who knows the smallest thing about arithmetic that the answer should be "7." Something not unrelated happens in the case of the conversion of my height in inches to my height in centimeters. I am 69 inches tall and there are 2.54 centimeters to the inch. Punch the sum into a calculator and the answer that comes out is "175.26." Yet to give my height in centimeters to two decimal places is plainly ridiculous: my height varies by more than .25 centimeters depending on how I brush my hair and it varies by much more than .01 centimeters every time I breath in and breath out. What counts as acceptable levels of approximation and precision is one of those things that only people embedded in the appropriate social groups can understand, and though computers can be built to almost any degree of precision or nonprecision, the choice of correct level of precision is beyond them. Nevertheless, we still consider calculators to be better arithmeticians than ourselves. (This argument is made in Collins, *Artificial Experts* [1990].)

place within an action tree, which is always embedded in actions dependent on tacit knowledge. In the terms introduced earlier in the book, mimeomorphic actions and mechanical causal sequences are embedded in action trees, which always start and mostly finish with polimorphic actions.[7] Once more, however long the string and however long the sequence of transformations that takes place in the calculator, no arithmetic will have been afforded unless the cultural gap between programmer and user is not too big—the calculator is no use to a tribesman from the Amazon jungle—at least not in the absence of condition 4 and condition 5 transfers.

In medicine, prostheses rarely work in exactly the same way as the part they replace. What happens is that the relevant elements of the embedding organism makes up for and metaphorically "repair" the difference between the original and the prosthesis. It is the same with the calculator. The calculator can only work as a *social prosthesis*, the deficiencies of which are made up for and repaired by the surrounding social organism. This is what is happening when the human user does the approximating—the human is

7. See Collins and Kusch 1998, passim.

repairing the deficiencies of the calculator by fitting its output in with social expectations.[8]

What if I use the record-and-playback method to train a machine to spray chairs with paint? Have I told the machine something explicit or does the machine now have the tacit knowledge of the trainer? A paint sprayer head can be fixed to the end of a multijointed robot arm, the movements of which can be monitored and recorded. The first part of the operation requires a skilled human to guide the spray head through a job, say, the complex maneuvers needed to spray paint an intricate metal chair. The movements of the spray head are recorded on magnetic tape. Henceforth, the spray head can be made to repeat these movements by replaying the tape so that it drives the motors, reproducing the human operator's movements. Provided that the chairs to be sprayed are all the same so that the operator would naturally spray the next chair, and the next chair, and so on with the same set of movements, such a device can be used to spray paint further chairs without the operator. In the case of, at least, early machines of this type, the string on the tape would be a series of magnetic signals of continuously varying strength—an analogue string. It is unlikely that the string will ever be examined by a human—the string never passes through the human mind. The string, if it was looked at directly (I am not quite sure how it would be looked at "directly") would be incomprehensible to a human—it is "inaccessible." What happens is that the machine transforms the magnetic impressions on the tape into string that affords meaning when it spray paints a chair: when the machine is spraying a chair the string becomes transformed in such a way as to become accessible. (Think of the paint sprayer as a kind of magnifying glass for the strings inscribed on the tape that drives it.)

The paint sprayer is a particularly revealing example. The human who sprays the chairs appears to have his or her movements transformed into strings by the machine. But to spray chairs, humans (assuming they are not trained after the fashion of galley slaves) have to understand what chair

8. An example of the way numbers sometimes need interpreting by the body (somatic tacit knowledge—see chapter 4) is the fireproof safe in my office. The instructions for opening it could certainly be chanted and might be though to be a paradigm of "explicit knowledge." They run, "start with a few clockwise turns, then make three turns counterclockwise until number X is opposite the pointer, then two turns clockwise until Y is opposite the pointer, then one turn counterclockwise until Z is opposite the pointer, then turn clockwise." It is remarkable how difficult this is to accomplish. There is a lot of interpretation involved in knowing what counts as "a turn" and how exact the stopping point at each turn has to be. A lot of manual dexterity is required to achieve enough accuracy without overshooting and of the numbers and novices try to correct a slight over-shooting by going back a little way, which is fatal. It takes everyone who has to use this safe a long time to learn the knack.

spraying means. They have to know that paint costs money and should not be wasted, they have to know what a satisfactory job looks like—how thick the paint should be on the surfaces, which are the important surfaces and which less important (which in turn depends on an understanding of how chairs are used), and so forth. So chair spraying depends on an understanding of human society, economics, and sitting.[9] Making sense of the spray head movements depends on the same social understandings. Are all these social understandings transferred to the strings on the magnetic tape? It is tempting to say so. The temptation is enhanced because the strings are generated, not by a team of programmers such as might program a pocket calculator in an *accessible* way, but by an *inaccessible* process called "training." The machine is "trained" to spray the chairs, rather than programmed.

As explained earlier, on the whole, digital strings are accessible. That is, it is possible to comprehend and describe every element of a digital string and the work it is doing. The way a digital string operates can be followed and the role of each of its elements can be described. For example, it is possible to describe the work done by the letter "A" in "cat" and, with a bit more work, it is possible to describe the work done by so-called computer languages (which are not really languages at all), even when transformed to the level of 0s and 1s. One can say exactly what would happen if one of the hundreds of millions of 0s in a long computer program was replaced with a 1. Indeed, the discovery that an unintended effect has been brought about by something like the accidental replacement of a 0 with 1 is the process called "debugging." The *accessibility* of digital strings—the fact that we can see and describe how they work—is one of the things that makes it tempting to refer to digital string as "explicit knowledge."

Analogue strings are not so accessible, and this tempts us to think of them as having something to do with the tacit. In our terminology, the *Mona Lisa* is an analogue string, or collection of strings (corresponding to eyes, lips, and so forth). You cannot, however, comprehend and describe the work done by the eyes and lips of the *Mona Lisa* in producing the smile. You cannot say, "if I replace this bit of paint with that bit of paint the *Mona Lisa* will smile better." The artist may accomplish such changes but you cannot say exactly how it is done. The *Mona Lisa* is an extreme example and I am really not sure whether the house/hut pictographs of chapter 2

9. Edgar Whitley pointed out to me that builders generally do not paint the top and bottom surfaces of doors, as these are never seen. In case of an unusual staircase location, however, such as would reveal the top of an open door to those descending, the top surface might be painted.

should be said to be accessible or inaccessible—after all, you can describe the work done by the strokes that make up the roof. But nothing turns on this ambiguity. What is important is the idea that some strings are accessible and some not and in ordinary discourse accessibility tends to be associated with the explicit while inaccessibility is associated with the tacit. Hence the record-and-playback chair sprayer seems, at first sight, to capture the tacit knowledge of the human who "trains" it, but more careful analysis shows that it is only doing string transformation that merges into mechanical cause and effect. By now this should be obvious, because we know that even an inaccessible analogue string can be turned into an accessible digital string with enough work.

The temptation not to keep this in-principle point in mind is nicely illustrated by some of the things that have been said about the computers known as "neural nets." The string of a neural net is intriguingly inaccessible and tempts commentators to think it is doing more than it is. The design of neural nets is inspired by the relationship between neurons in the human brain, where it is thought that the pattern of connections is continually reforming itself and is the seat of human knowledge. The neurons are linked via connections of varying strength; a strong connection is likely to make the connected neuron "fire" when the first neuron fires, whereas a weak connection makes this happen less often. That is the way we build up associations between things in the world, the basis of our understanding. Neural nets have nodes linked by connections that take on different strengths according to whether a joint firing is reinforced or discouraged. The neural net, then, like the paint sprayer, is not so much programmed as it is trained.

Thus, suppose I am training a neural net to speak written words. To caricature, I type in the word "cat" and the net shuffles its connections randomly and utters a sound; I keep doing this until the net utters a random sound that is a bit like "cat." I press the reinforcement button, as it were, and the connections between the typed word and the sound that has just been uttered are reinforced a little. Henceforward when I type "cat," the random sounds emitted all sound a little more like "cat" than before and if I wait until I hear a sound that is still more like "cat" and reinforce that, I can refine the net's connection weights still further. By a process of iteration like this (which is likely to be automated), the net will eventually learn to speak an indefinite number of typed-in words—the net will have learned to speak in the sense of putting the right phonemes together with the right written words.

Neural nets gave rise to a lot of excitement when they were first being developed and they still do.[10] Neural nets seemed to learn like human beings rather than having to be programmed like computers. That the "learning" was, at best, a simple kind of Skinnerian conditioning of an isolated individual that is nothing like the process through which humans learn languages did not seem to deter enthusiasts. I suspect this was because, unlike previous generations of "intelligent" computers, the strings of neural nets are inaccessible—just like the strings used by humans! Because their strings are inaccessible, neural nets seem to be different to plain old digital computers. With plain old digital computers the strings have to be accessible because someone has to write them. With neural nets the machine seemed to write its own strings as it interacted with the people around it—again, like us. But, of course, the strings of a trained neural net *are* accessible. First, every string is accessible in principle, though we know that there may be almost insurmountable logistic barriers to accessing it. In the case of neural nets, however, the barriers are not insurmountable because nothing as complicated as reducing everything to atoms or quantum states is necessary, since all the nets that have been instantiated on ordinary digital computers. Therefore, take any such computer apart and there would be the strings, in the form of 0s and 1s, ready to transport without loss to any other computer, which would thereby speak in a way identical to the first one—quite unlike what we can do with human beings.[11]

When we write a program for a digital computer, all of it—even if it is only a bit at a time—has to be accessible to someone at some time. Consequently, we say that the program is explicable. This is one of the rea-

10. See Powers 2004.

11. Another kind of seductive machine is speech transcriber, which has to be "trained" by speaking into it a passage of text that it already "knows." Speech transcribers take the string of regularly occurring sounds in a person's speech, transform them into patterns of electrical charges, and then transform them again into letter shapes on a screen. The art is in the choice of boundaries to the sound classification scheme so that what counts as "regularly occurring" to the machine has some correspondence with "regularly occurring" in a particular person's speech. From the moment the words are turned into sounds, of course, meaning is absent—sounds and the inscribed words they are transformed into are just strings. The meaning has to be reimposed by the reader, with luck, without too much need for correction. When the speaker speaks very carefully in the attempt to get the machine to work properly, the first step of turning a language into a string has been taken: changing ordinary speech into something more like a string. Speech transcribers are really "sound transcribers." As they become more powerful, more and more of the transformation will be able to be done by the machine, with less and less help from the human. It may even be that the amount of "training" that is necessary to get the machine to associate certain electrical strings with certain sounds specific to the user will be reduced. There is no mystery to any of this. It is just like spraying letters on a crate using a template: the speech transcriber is just a complicated template.

sons why computers have so changed the debate about what tacit and explicit mean. The reason for celebrating the invention of neural nets (and it should have been a reason for celebrating record-and-playback machines such as paint sprayers too) was that they seemed to do something tacit. But the celebration, if there was to be one, should have been about how something tacit was being made explicit, not how a machine was acquiring tacit knowledge. In truth, there should have been no celebration at all, because the paint sprayer can only spray one kind of chair, not chairs as a whole, and the neural net can do no more than the equivalent of Pavlov's dog. Both are simply string transformers the operation of which merges into the ordinary world of mechanical cause and effect.

Does a sieve have the knowledge to sort big items from small items and, if "yes," did the designer "tell" the sieve how to do it? Does my cat have tacit knowledge of how to hunt? It doesn't have explicit knowledge! In the introduction it was claimed that a large part of what counts as tacit knowledge is quite unexceptionable because human use of it is it is continuous with what cats, dogs, and trees do. Polanyi himself made the point:

> The ineffable domain of skilful knowing is continuous in its inarticulateness with the knowledge possessed by animals and infants. . . . The anatomist exploring by dissection a complex topography is in fact using his intelligence very much like a rat running a maze. . . . We may say in general that by acquiring a skill, whether muscular or intellectual, we achieve an understanding which we cannot put into words and which is continuous with the inarticulate faculties of animals.[12]

Of cats, dogs, and trees we can certainly state that given a generous interpretation of the word "know," like the surgeon described by Polanyi, they know more than they can tell. Should we, then, say that cats, dogs, and trees have tacit knowledge? And while we are pursuing this line of inquiry, should I say that a sieve has tacit knowledge in that it "knows" how to sort gravel but it too cannot tell how to do it?

The reasons for saying that animals do not have tacit knowledge is that they do not keep secrets and they do not encounter the problem of rules not containing the rules for their own application (the rules regress), because they never do any interpreting of strings and therefore never come up against the limits of interpretation. And, of course, they do not share collective tacit knowledge of the kind that will be discussed in chapter 6. One

12. Polanyi 1958, 90. This quotation also serves as the epigraph to chapter 5.

might claim that they learn, at least partially, only through direct contact with other animals, but as has been argued, and as we have seen with neural nets and the paint sprayer, this does not necessarily mean that anything is going that is not explicable in one or more senses of the word.

What a tree and a sieve do has already been described—they are mechanisms that, on our analysis, merge into string transformers.[13] Apart from the fact that sieves and trees do not have brains, whereas cats and dogs do have them, there seems no reason not to think of the latter as the same kind of entity. What difference does the brain make? The only difference is that cats and dogs can more easily be subject to condition 4 transfers. Their substance can be a little more easily changed than is the case with sieves and by a different process than is the case with computers. Sieves and trees do only one thing whereas cats and, more obviously, dogs can be transformed by training rather than physical implants in such as way as to retrieve dead pheasants and the like. The training, however, is conditioning, or shaping behavior, and almost certainly does amount to a fairly inflexible physical transformation the brain.[14]

It has already been anticipated that I am going to claim in chapter 5 that humans, *insofar as humans act as animals*, are not so different from cats and dogs. If we are going to say that humans have tacit knowledge when they act like animals (as on Polanyi's usage), this does not rule out cats, dogs, trees, and sieves having tacit knowledge. Indeed, Polanyi's correct insight about the continuity of animals and humans (as animals), which is expressed in a rather different way in the quotation on page 76, provides a useful "handle" on what this kind of tacit knowledge is like. When I want to think about the embodiment or "embrainment" of some kinds of human knowledge I need only think of cats and dogs—or trees and sieves for that matter.

The problem is a bit like that of the second description of rugby ball kicking—the description in terms of coordinate geometry. In that case it was concluded that the scientific description was a good candidate for being called explicit knowledge even though the human could not use it. It

13. It might tempting to think that the sieve is a digital device, as the stones either fall through or don't so that there are no in-between states that cannot be examined. More careful consideration of exactly how a sieve works will reveal, however, that which stones fall through is a very complicated function of the shape of the holes in the sieve, the shape of the stones (not just the longest dimension of the minimum cross-section, as might be thought), their coefficients of friction, and the duration and vigor of the shaking.

14. I have never understood the work that the idea of "consciousness" is supposed to do in this kind of discussion, though many others seem to think it important.

seems the same with animals, trees, and so forth. We can provide a scientific explanation or description of what they do so the knowledge is explicable in that sense. But it is explicable only for us, not for animals and things. Thus, the animals, sieves, and trees seem at first sight to know more than they can tell. One might, then, want to say that they have tacit knowledge even though all they are doing, in our terms, is string transformation—mostly transformations of very complex analogue strings.

Nevertheless, as argued in the introduction, the idea of tacit knowledge only makes sense when it is in tension with explicit knowledge, and since cats and dogs and sieves and trees cannot be said to "know" any explicit knowledge, they shouldn't be said to know any tacit knowledge either. In fact, they don't "know" anything; they just transform strings. Cats, dogs, trees, and sieves just hunt, sniff, grow, and sift in the way that a river flows. To borrow the Wittgensteinian sentiment, to say that a tree knows how to be a tree or that a cat knows how to hunt (or that a calculator knows how to calculate) is to invite language to go on holiday.[15] Such language should be kept for children's stories.[16] In sum, animals, trees, and sieves should not be said to have tacit knowledge.

What if I have a special grip on my golf club that ensures that my hand assumes the right position? Is the special grip telling my hand what to do? The special grip on my golf club is like the uninterpreted forceful push used for moving a spectator to one side. The grip physically pushes my hand to the right position. I might then play my shot but, of course, the point of the grip is for me to take note of the proper hand position and resume it even when the grip is no longer there. So the grip affords a meaning "this is how to hold a golf club." Seen this way it is an example of analogue string. Does it tell my hand what to do? Yes, in the vernacular, but "analogue string that affords the meaning 'grip the club this way'" is the right way to talk about it.

What if I can write out the mechanical formula for balancing on a bike? Does that mean that bike riding is explicable? Writing out the mechanical formula for balancing on a bike is like describing the flight of a rugby ball in ballistic terminology in terms of coordinate geometry. It is a string that might afford enough to be useful in general knowledge quizzes and it might be transformed so as to have a function in mechanized bike balancers. More will be said about bike balancing in chapter 5.

15. If it were to turn out that the higher primates and dolphins could be shown to have languages rather than strings then this argument might not apply to them in quite such a clear-cut way.

16. Or actor-network theory, as in Callon 1986.

(4) Questions about Coaching and Human Animals

What if I tell a novice that he or she should look well ahead, not down at the ground, when trying to learn to ride a bike? Is that telling the novice how to ride explicitly? Telling a novice to look well ahead when learning to ride a bike is like the first description of rugby ball kicking (found in box 2). It is a string that, beyond general knowledge answers, can help a novice who knows enough to understand the words and has the muscular development, confidence, and bike to put it into practice. It is a coaching rule, and coaching rules are one form of string that affords the development of a practice. If one wants to talk in ordinary speech of it being explicit knowledge no great harm is done so long as false inferences are not drawn from this everyday usage.

When a person has learned to kick a rugby ball or balance on a bike, the person has acquired what Polanyi would call new pieces of tacit knowledge. He would call them this because the person can only partially describe what is done to accomplish these physical feats and the description, however full, would not in itself enable the hearer to do the same thing. At best the hearer could use the description as coaching rules or second-order rules to assist with the mastery of the tacit knowledge, which is located in brain connections, nerves, and muscles as much as in memory. I have intimated, and will press the point later in the book, that everything that is done by the rugby ball kicker and bike rider (if by "riding" we mean balancing on the bike while it moves along), could, in principle, be represented by strings of one sort of another—in other words they are fully "explicable" in the scientific sense of "explainable." Nevertheless, as we will see in chapter 5, it is conventional to refer to this kind of ability in the way Polanyi and many other writers do, as tacit knowledge. If we had a question about whether to call it tacit or explicit knowledge, it would be not so different to the question about whether we should call what cats, dogs and trees do tacit or explicit. Here, again, the crucial thing is not to mistake the everyday language of tacit and explicit with the technical language of strings and languages.

If I act for reasons that are subconscious, are they tacit and do they become explicit if the psychiatrist uncovers them? The psychiatrist, assuming psychiatry has some value, is making something explicable and so rendering something explicit. But it is not a deep sense of explicable since, presumably, my unconscious drives are just instances of the mechanisms that drive me. The hard part of this is the initial repertoire of drives that I might have. If this repertoire of drives is partly provided by the way I am embedded in society then the psychiatrist will not know how to make explicit how that repertoire was initially formed.

Conclusion to Chapters 1–3: What Are Tacit and Explicit Knowledge?

In this chapter the terms developed in chapter 1 and analyzed further in chapter 2 have been used to find a way through a series of puzzles about the everyday use of the words "tacit" and "explicit" in academic life. We are now in a position to explain the tacit and the explicit. That which is not explicit knowledge is mostly just the way the world unfolds. Sometimes it is referred to as tacit knowledge. Much of it consists of the working out of mechanical sequences of greater (cats, dogs, humans-as-animals, paint sprayers, neural nets) or lesser (trees, sieves) complexity. For most of this the term "tacit knowledge" should not be used, the notion of mechanism being more appropriate. It has also been argued that strings affect entities in four ways, and the first three of these also merge into good old-fashioned causes and effects. So both tacit and the majority of explicit communications are just mechanisms. The fourth way in which strings affect entities is the only exception to this—interpretation is truly interesting. In the other three cases of strings affecting entities, what makes them worthy of remark is that the causal effect is dominated by the string pattern whereas with tacit knowledge we don't see much of the pattern.

The reason we do have a number of worthwhile if incomplete analyses of tacit knowledge per se, is that the modern world is thought of as driven by explicit knowledge—patterns. The explicit is taken as the norm rather than the tacit—and the contrast with what is not explicit is ever present. In other words, as soon as we begin to reflect and write our reflections down we create a subject which is concerned with what cannot be written down. That is why there is a topic worth expanding over the next three chapters in which it will be claimed that there are three kinds of tacit knowledge. These are, then, three main kinds of reason for not being able to write things down.

There is no animal explicit knowledge and, consequently, no contrast that would make sense of the term tacit knowledge in respect of animals. So in this sense, "explicit" is a relative term. Nevertheless, explicit knowledge has substance—it is knowledge that can, to some extent, be transferred by the use of strings in the right circumstances. The circumstances are dealt with in the most general terms under the heading of the conditions of communication.

Table 4. Four meanings of "explicable."

1. Explicable by elaboration	A longer string affords meaning when a short one does not.
2. Explicable by transformation	Physical transformation of strings enhances their causal effect and affordance.
3. Explicable as mechanization	A string is transformed into mechanical causes and effects that mimic human action.
4. Explicable as explanation	Mechanical causes and affects are transformed into strings called scientific explanations.

When we use the term explicit we draw on a subset of the ways we talk about communication. We don't count enabling conditions 4 and 5 of communication as "rendering explicable," because they comprise changes in the receiving entity rather than changes in the string. That is to say, even though a string that initially cannot do work can be made to do work by physical changes in the entity upon which it impacts, we do not say that these changes render the string explicit: this is just how we use words. The four ways we actually use the term explicable are now set out in table 4.

This concludes the analysis of explicit knowledge. We can now move on to tacit knowledge. The next three chapters can be thought of as exploring further the articulation between the conditions of communication and table 4.

PART II

Tacit Knowledge

FOUR

Relational Tacit Knowledge

In the first three chapters the ideas of the explicit was reformulated in terms of strings and things. The intention was to make it possible to embark on an exploration of the notion of the tacit that is less likely to stumble over conceptual difficulties and ambiguities. In the introduction it was argued that the notion of the tacit was parasitic on the notion of the explicit. If it were not for the idea of the explicit we would never have noticed that there was anything special about the tacit—it would just be normal life. Having invented the explicit, however, we now have the tacit. The tacit is that which has not or cannot be made explicit. Accordingly, the next three chapters analyze weak, medium, and strong tacit knowledge—these adjectives referring to the degree of resistance of the tacit knowledge to being made explicit. Working backwards, chapter 6 deals with strong tacit knowledge, otherwise known as collective tacit knowledge (CTK). It is a kind of knowledge that we do not know how to make explicit and that we cannot foresee how to explicate in any of the senses of explicable offered at the end of the last chapter. Strong tacit knowledge, as intimated, is the domain of knowledge that is located in society—it has to do with *the way society is constituted.* Among other things, chapter 6 describes the changes that take place when condition 5 communication is satisfied.

Medium tacit knowledge, or somatic tacit knowledge (STK), is dealt with in chapter 5. It has to do with properties of individuals' bodies and brains as physical things. I have intimated and will argue further, that this kind of tacit knowledge is continuous with that possessed by animals and other living things. In principle it is possible for it to be explicated, not by the animals and trees themselves (or the particular humans who embody it), but as the outcome of research done by human scientists (the third meaning of explicable). It is possible to foresee, therefore, that we will one day be able to mimic animal behavior with machines. The same goes for humans-as-animals, though it will probably take a lot longer. What this means for the concept of tacit knowledge is explained in the chapter. Chapter 5 explains, among other things, what happens when condition 4 communication takes place.

Weak, or relational, tacit knowledge (RTK), is the subject of this chapter. It is knowledge that could be made explicit in the second sense of explicable but is not made explicit for reasons that touch on no deep principles that have to do with either the nature and location of knowledge or the way humans are made. Collective tacit knowledge turns on the nature of the social, somatic tacit knowledge turns on the nature of the body, but relational tacit knowledge is just a matter of how particular people relate to each other—either because of their individual propensities or those they acquire from the local social groups to which they belong. Relational tacit knowledge turns on *the way societies are organized*. In the case of weak tacit knowledge, both sender and receiver already have enough cultural similarity for a string to afford the intended meaning to the receiver if the string was long enough, but the sender either feels no inclination to make the string long enough or does not know how to make it long enough. In short, what are dealt with in this chapter is condition 1 transfers that fail but that could work as condition 3 transfers like the numbered jokes in the pub. The subject of the chapter is why the condition 3 transfers do not happen. This is a very simple idea. Weak, or relational, tacit knowledge is the first phase of the Three Phase Model of tacit knowledge, the phases being, of course, RTK, STK, and CTK.

The three chapters, *inter alia*, explain conditions 3, 4, and 5 of knowledge transfer, respectively. They explain why condition 3 does not always work and, in chapter 6, it will be explained why in certain cases it will not work in the foreseeable future—we cannot see how to make it work in the realm of the social.

Preliminaries

Before these chapters on tacit knowledge can be properly begun, two pieces of preliminary work need to be completed. The first of these is to describe the difference between the ways we encounter explicit and tacit knowledge; and the second, given that tacit knowledge is that which cannot be or has not been made explicit, is to say what is meant by "cannot."

Communication with and without Middle Persons and Things

The explicit has to do with the transmission of something via strings from which it follows that the tacit cannot be or is not transmitted with strings.

The explicit, then, can be conveyed by middle persons or middle things with strings inscribed upon them; the tacit must involved direct contact. Thus, if a middle person appears to transmit the tacit it cannot be in the form of a string inscribed upon them; it must be in some other form. They must, therefore, already own whatever it is—they must, as it were, already be an end person, not a middle person. The tacit is communicated by "hanging around" with such persons. In children and older students tacit knowledge is acquired by socialization among parents, teachers, and peers. In the workplace it is acquired by "sitting by Nellie" or more organized apprenticeship. In science it is acquired during research degrees, by talk at conferences, by laboratory visits, and in the coffee bar.

It is important to note that there is an asymmetry in the matter of intermediaries and knowledge. One can say for sure that knowledge that can be passed on via intermediaries is explicit knowledge, but one cannot say that knowledge that is passed on without intermediaries is tacit knowledge. Close personal contact between teacher and learner enhances the transmission of explicit knowledge as well as tacit knowledge, so close contact proves nothing about the nature of the knowledge. Furthermore, certain potentially explicable forms of tacit knowledge are so complex that the only practical way to transfer them from human to human is by the close personal contact that allows for guiding, showing, imitating, and so forth as a short cut to explaining. The fact that *some knowledge* that is transferred by close contact is explicit knowledge, and therefore could have been communicated by intermediaries, can lead to the false inference that *all knowledge* that is transferred by close contact could be communicated by intermediaries. It is the same kind of false inference that is drawn from the fact that sometimes longer strings will work when shorter strings will not or that sometimes a computer will enable a string to be read or used to produce a mechanical output when humans beings could not make use of such strings (so it is inferred that robots that can replace us are just round the corner). It is the sort of mistake that leads to the false inference that, with enough work, an entire education can be transmitted via intermediaries such as the Internet. But this is a terrible mistake; education is more a matter of socialization into tacit ways of thinking and doing than transferring explicit information or instructions.[1]

1. See Dreyfus 2001.

The Meaning of "Cannot"

The word "cannot" has already appeared a number of times in this text. Different people mean different things by "cannot," and this makes it hard to argue the case that there is anything that cannot be done. Furthermore, saying that something cannot be done sounds like prophecy and it is a regular feature of human experience that prophecies (humans cannot travel faster than 30 miles per hour, heavier-than-air flight will never be achieved by humans, no human will ever leave the Earth's atmosphere or gravitational field) are often confounded by technological breakthroughs. In the next chapters, however, the question is going to arise about which kinds of tacit knowledge can be made explicit under one meaning of explicit or another. Therefore the term "cannot" must be defined in a way that reduces its ambiguity insofar as this can be done. This exercise has to be carried out somewhere and this is as good a place as any.[2]

In table 5 a series of decreasingly significant meanings for "cannot" is set out. Henceforward, "cannot," when used "in anger," in this text, will usually be accompanied by one of the qualifiers found in the first column of table 5. The second column of the table illustrates the meanings with sentence-length examples.

Cannot (*logical impossibility*) is of no interest in this book, but the entry is included so that the table contains as exhaustive a list of "cannots" as possible.

In this book it will not be proved that anything cannot be done in the sense of **cannot (*scientific principle*)**. This is not to say that some of things that are discussed might not turn out to be impossible in principle. After all, the impossibility of faster-than-light-speed travel, like the impossibility of perpetual motion machines, was discovered quite late in the history of the respective ideas. When it is claimed, as it will be claimed, that something cannot be done by any *foreseeable* means it certainly leaves open the possibility that the reason is one of deep principle that we do not yet know. Thus, some time in, say, the eighteenth century, someone might have claimed, "there is no foreseeable way to make a perpetual motion machine," without knowing that in due course the principle of conservation of energy would show that it would always be unforeseeable. It may be that there are such principles still to be uncovered in the case of the nature of knowledge, but here I cannot claim to be discovering such principles. When I make a claim of the kind "I cannot foresee how this kind of tacit knowledge could

2. Barrow (1999) has also attempted to study the meaning of the impossible.

Table 5. Types of "cannot."

Cannot (*logical impossibility*)	We cannot have our cake and eat it too.
Cannot (*scientific principle*)	We cannot travel faster than light.
Cannot (*logistic principle*)	We cannot enumerate and store all possible chess games.
Cannot (*logistic practice*)	We cannot build a rocket that will accelerate a human to a speed of 280,000 kilometers per second.
Cannot (*technological impossibility*)	We cannot hold a conversation with anyone more than a mile away. / We cannot make rechargeable car batteries with the energy density of a tank of hydrocarbon fuel.
Cannot (*technical competence*)	We cannot translate the Rosetta Stone. / We cannot make room-temperature superconductors.
Cannot (*somatic limit*)	We cannot run a mile in less than two minutes.
Cannot (*contingency*)	I cannot tell you because it's a secret. / I cannot tell you because I do not know what you need to know.

become explicit," at best I remain agnostic in respect of the possibility that, on the one hand, scientific principles might one day be found that forbid it in perpetuity (or in the same sense that faster-than-light-speed travel is forbidden) or that, on the other hand, what cannot now be foreseen will one day be accomplished by so far unimagined means.

Cannot (*logistic principle*) is a little less absolute than cannot (*scientific principle*), but not much. We do know a process for enumerating and storing all possible chess games (of a limited length) but we know it would need a memory store bigger than our universe.[3] Cannot (*logistic principle*) will again play only a small role in this book, but it remains a possibility that something that seems impossible for some lesser reason will turn out to be impossible for this greater reason.

We know how to build a rocket that can accelerate a human to 280,000 kilometers per second and will not use up more resources than there are in the universe, but it would require so many resources that we cannot imagine actually doing it. **Cannot (*logistic practice*)** will play an important role in the book as an obstacle to the complete explication of tacit knowledge.

The first example of the use of **cannot (*technological impossibility*)** given in table 5 is, of course, wrong. For example, we can hold transatlantic conversations via the telephone. The point is that this first illustration would

3. Haugeland (1986, 178) writes that to store all chess games of forty moves in length would require more memory locations (10^{120}) than there have been quantum transitions in the universe. Regardless of whether this is exactly right, the idea will do for the purpose at hand.

have been believed by most people before the invention of the telephone or radio transmission (or smoke signals, if you insist). It is an example of a use of cannot that, prior to telephone or radio transmission, would have appeared in the following sentence: "We cannot hold a conversation with anyone more than a mile away unless some new and currently unforeseeable principle is invented." The second illustration of cannot (*technological impossibility*) can still be correctly stated today: "We cannot make rechargeable car batteries with the energy density of a tank of hydrocarbon fuel unless some new and currently unforeseeable principle is invented." Cannot (*technological impossibility*) is going to play a big part in this book.

The first and second illustrations of the use of **cannot (*technical competence*)** are related in the same way as the first and second illustrations of the previous cannot. At one point the Rosetta Stone could not be translated, but those who were trying to translate it would have believed—correctly, as it turns out—that no new principles were needed to complete the task. It looks as though it is going to be the same with superconductors. Superconductors that work at higher and higher temperatures are regularly being made with new mixtures of well-known compounds, and it looks as though room-temperature is being approached. (This, it should be understood, is not a prediction about superconductivity, it is an illustration of the use of the word "cannot.") There are going to be people who will argue that my usages of cannot (*technological impossibility*) in this book are really examples of cannot (*technical competence*). I think I am right—they think they are right. Nothing of huge importance to the understanding of tacit knowledge turns on the matter, however. What does turn on it is the future direction of artificial intelligence research; I believe it is misdirected but that point has been argued elsewhere.[4]

As can be seen, sometimes lower levels of "cannot" don't exclude higher levels. It has just been suggested that some people will say that the higher level, *technological impossibility*, is really mere *technical competence* but they can't be absolutely sure of this. Indeed, it may be that there is some as yet entirely unknown reason of scientific principle that will always prevent room-temperature superconductors being made. To fantasize, perhaps the lossless transmission of energy across the surface of a planet fit for human life is impossible due to some principle related to thermodynamics. What we do here is use the highest level of cannot that can be proved. The highest level of cannot that can be proved is usually a fairly low level, which is why the strongest claims in this book are to do with *technical competence* (with the occasional invocation of *logistic practice*). In other words, when a low

4. For example, in Collins and Kusch 1998.

level impossibility claim is being made it is often compatible with a higher level impossibility claim that cannot be proved or disproved. Occasionally this will be indicated by including "or higher" in the bracketed qualifier as in cannot (*technical competence or higher*).

Cannot (*somatic limit*) will also be used a lot in this book. It is about the limitations of human performance. I doubt that these usages will be contentious.

Cannot (*contingency*) will also play a role in the book. The "weakest" form of tacit knowledge—relational tacit knowledge—are cases where the parties could tell each other what they need to know but either will not, or cannot for reasons that are not very profound, such as not knowing what the other party needs to know.

Relational Tacit Knowledge

Concealed Knowledge

Sometimes knowledge that can be told—conveyed with a few words—is deliberately kept hidden, as in the passage quoted in box 7 that refers to the transmission of information among laser scientists.

Box 7. Concealed knowledge.

> One scientist reported the following of a visit to another laboratory:
> "They showed me roughly what it looked like but they wouldn't show me anything as to how they managed to damage mirrors. I had not a rebuff, but they were very cautious."
>
> A more subtle tactic used was that of answering question but not actually volunteering information. This maintains the appearance of openness while many important items of information are withheld; their significance will not occur to the questioner. One scientist put it:
> "If someone comes here to look at the laser the normal approach is to answer their question but . . . although it is in our interests to answer their question in an information exchange, we don't give our liberty."
>
> Another remarked succinctly:
> "Let's say I've always told the truth, nothing but the truth, but not the whole truth" (Collins, *Changing Order* [1985], 55)

Is there any reason to refer to the content of box 7 as tacit knowledge? Why is it not just a matter of keeping a "secret"? This question is not deep and should not be treated as a philosophical puzzle; we know that the term "tacit knowledge" is used in many ways and often used carelessly. In this chapter secrets of this kind will be grouped in with other types of tacit knowledge. Why this choice? First, note that the secrets are not of the same kind as those in the coded U-boat rendezvous message or even in the untranslated Rosetta Stone. They differ in that they are not written down so they are, as it were, less explicit than the other two cases. Second, the knowledge often can be gained by the normal means of transference of tacit knowledge—close contact with the owner of the knowledge in the absence of intermediaries; the cunning visitor to a rival scientist's laboratory can gain a lot just from looking closely at what is being done even if the other scientist will not spell it out. Third, secrets of this type are continuous with other closely related bits of knowledge, which are also not transferred for not very deep reasons. All of these are bits of knowledge that are not transferred simply because the strings that pass between the sender and receiver are not long enough to afford what the receiver needs to know in order to act in the new way. That is, all these kinds of knowledge are not explicit knowledge so we might as well call them "tacit." It is a ragbag category—but a vital one. These piece of relational tacit knowledge are, of course, *explicable* in the first sense of the term, but that just reconfirms the confused nature of the existing vocabulary.

Knowledge is also sometimes kept secret by elite groups. Secret societies have certain rituals that are kept hidden so as to exclude outsiders. But these secrets can acquired by infiltrating the group and seeing or experiencing their rituals firsthand. Indeed, this is the whole point in these cases: the knowledge is vouchsafed to those within the appropriate social network or social space—who can communicate with "a nod and a wink"—but hidden from those who do not belong to the in-group. The very idea is to demarcate those who have spent time in the right networks from those who have not. Or, in the case of apprenticeship, when tricks of the trade will not be told because humiliation of the uninitiated seems part of the ritual and the power relations, the tricks will eventually be picked up through watching and trying to copy the master (but not before a lot of cheap labor has been extracted from the apprentice; see box 8).

Some of the knowledge that is learned through joining the networks of elite groups or getting close to masters is knowledge that could be told, and some is not. We are still on the borderline of the explicit and the tacit.

Box 8. Japanese apprentices and concealed knowledge.

"It is expected that serious learning will proceed unmediated by didactic instruction. . . . When an apprentice presumes to ask the master a question, he will be asked why he has not been watching the potter at work, where the answer would be obvious. . . . Learning of a craft skill is often described in Japan as 'stealing the master's secrets.' Numerous apocryphal stories are told of enterprising apprentices who find stealthy ways of discovering the secrets of their master's craft production . . . The master is seen as protecting his 'secrets' and the learner is expected to 'steal' them. (John Singleton, "Japanese Folkcraft Pottery Apprenticeship: Cultural Patterns of an Educational Institution," in *Apprenticeship: From Theory to Methods and Back Again*, ed. Michael Coy [Albany: State University of New York Press, 1989], 26)

Ostensive Knowledge

Another part of the domain of the tacit as defined in this chapter is *ostensive knowledge*—knowledge that can be learned only by pointing to some object or practice because the description in words, though everything is there to see and be described, would be too complex to be spoken and apprehended.

Of course, according to chapter 2, even artifacts themselves are strings so if knowledge that could not be transferred with words, can be transferred by looking at an artifact then, it is a case of condition 3 transfer. So, in logic, once more, this might be said to be explicit rather than tacit knowledge. An artifact, like a picture is, as we might say, worth a thousand words (or many more).

But, again, the ordinary usage of "tacit" is sufficiently imprecise to give us the license to call it tacit knowledge in spite of what we know is going on when we think in terms of strings and entities. The feeling of discomfort can be relieved just a little by imagining that to enable the knowledge to be transferred the learner has ask questions, touch, walk round the object to get different views at different times, request that the object or apparatus be subjected to one kind of manipulation or another, or perhaps taken apart so that hidden parts of it can be seen. In these circumstances the artifact begins to feel less like an intermediary than an object in a social setting that involves personal contact and a bit more of the "color" of tacit knowledge seems to be present.[5]

Logistically Demanding Knowledge

Imagine a traditional manufacturing company with an old warehouseman who knows where everything is. If you asked the warehouseman to list the location of everything in the store by name, he would not be able to do it. But if you ask him for a specific part, perhaps describing its function or its shape and size, or showing him a broken version of what you want, something in his brain and, perhaps body, will bring him to the place where it is kept. The warehouseman has Somatic tacit knowledge of the warehouse, a category that will be discussed at greater length in the next chapter.

But thinking in terms of knowledge as something that enables a task to be carried out, the warehouseman's knowledge could be made explicit. The warehouseman could be replaced with a computerized system so long as someone devised a numbered set of racks and a catalog with enough classifications of the objects on the shelves to make them easy to recognize and match with other objects. The system would not do the job in just the same way as the warehouseman, but it could accomplish the same task when used as a social prosthesis. This solution would turn on condition 3 of communication—it would be a matter of creating a long string where only short strings were needed before.

If the warehouseman is reliable, however, it may be more efficient to keep things as they are. Organizations, so the literature goes, have to make choices like this all the time.[6] At what point do they invest in formal systems to take over the hands-on retrieval skills of faithful employees? As

5. Ribeiro (2007) considers proximity to artifacts and processes—what he calls "physical contiguity"—to be of special importance in the transfer of knowledge. According to the account given here, there is no conceptual discontinuity between physical contiguity and linguistic immersion in the absence of the artifacts. The only discontinuity occurs between attempts to gain knowledge without social contact and attempts to gain knowledge that do involve social contact. Physical contiguity is likely to shorten the time taken to gain understanding but it does not provide a different kind of understanding. That said, there are individuals, such as myself, who learn easier from artifacts than written texts. I have never felt comfortable doing sociology studies of fields that are not experimental and thus provide apparatus (or diagrams of apparatus) for me to look at; no doubt others learn better from texts and prefer to do studies in more theoretical areas. Ribeiro's discussion of the role of what he calls "different levels of immersion" is, therefore, of great practical importance.

6. Hedesstrom and Whitley (2000) discuss this point also making a useful distinction between those who treat the tacitness of knowledge as a matter of difficulty and those who treat it as a matter of choice about where to direct one's attention.

businesses grow larger and management becomes less paternal and workers less faithful to ever more-distant bureaucracies, much that depended on the tacit tends to become explicit—but not all of it. Again, the choice of which way to go—stick with the tacit or go to explicit—is a contingency of time and place so the knowledge, if it is kept tacit, is relational tacit knowledge.

Mismatched Saliences

Sometimes knowledge that can be told is kept hidden without it being anyone's intention to hide it. This happens when person A, who wants to convey everything they know to person B, assumes that person B is in possession of some essential piece of explicable knowledge (in the first sense) when in fact they are not. Since the transmission of even explicable knowledge can involve an indefinite number of pieces of information, A cannot resolve the problem by telling "everything" to B. A has to have a model of what is in B's head in order to make a stab at instructing B by filling in the missing gaps in B's knowledge, and A's model of the contents of B's head can be wrong. It is a common experience for a frustrated person to finish explaining by saying "Oh—I did not realize you did not know *that* [essential but trivial piece of information]!" In this case what seems salient in terms of explanation for A is not salient for B, so this is a case of "mismatched saliences" (see box 9).

The case of mismatched saliences might be said to be a stronger case of relational tacit knowledge than the previous examples, because however hard the teller tries to tell all, he or she cannot do it. While one can see that the reasons the knowledge is not made explicit are entirely matters of accident, the accidents cannot be avoided, so no one can volunteer to tell the knowledge. In such circumstances, however, providers of knowledge welcome close proximity between themselves and learners so that they can learn by every kind of interaction.

Unrecognized Knowledge

It may also be that A carries out certain procedures in certain ways but cannot tell B about it, because A does not know that certain of the ways in which they themselves do things are important—if they did know they could describe the important things. It sometimes happens, however, that A is just lucky enough to stumble on the right way of doing things and continues to do things that way because there is no reason to change. B is never going to be told about such things by A, because neither A nor B knows

Box 9. Features of tacit knowledge including mismatched saliences.

Something similar applies to the material of the suspension fibers. Checkhov used very fine Chinese silk thread, which he supplied to the Glasgow group (who had earlier used steel piano wire). Trial and error had shown the Russians that other kinds of silk thread gave lower Q's. It was also known that fine tungsten wire gave still better results, but that it had to be polished carefully to just the right (indescribable) degree and that the clamping problem was particularly acute with tungsten. Donald believed it was the hardness of the tungsten that made the clamping so critical—the compressibility of silk allowed a certain leeway in the design of the clamp. Thus silk was used for most runs, with tungsten (which might improve the Q by a factor of 2), being preserved for a final measurement once the general area of the expected result had been defined by the easier method. The nature of suspension materials and clamping . . . are matters whose salience became clear for the Glasgow group only after working with Checkhov. For both parties the science was slowly emerging and turning knowledge that no one knew they could or should express into something that could be articulated as the importance of previously unnoticed parts of the procedure became revealed. (From Collins, "Tacit Knowledge, Trust, and the Q of Sapphire" [2001])

that they are worth telling. This is a case of "unrecognized knowledge." It is quite possible that the crucial discovery will be made some time (see box 10) and then what was a piece of unknown knowledge will become known and capable of being told (condition 3).[7]

In the case of unrecognized knowledge, the tradition or habit in which the necessary knowledge is embedded might well be picked up by the matter-of-course imitation that close proximity to another is likely to engender. This, again, is how tacit knowledge is transferred.

To reiterate, concealed knowledge—knowledge subject to mismatched saliences and cases of unrecognized knowledge—are experienced by humans as tacit knowledge and acquired as tacit knowledge, even though it is not the "ontology" of the knowledge, nor even the structure of the human body and brain that have made them transferable in this way only.

7. "Unknown knowledge" is, of course, an oxymoron but not one that causes much trouble.

Box 10. Unrecognized knowledge becomes recognized.

"I worked with Dr. Bob Harrison as he set about building his first TEA laser. Among the things he knew was that the leads from the top capacitors to the top electrodes had to be 'short.' . . . It turns out that he did not interpret this knowledge properly and consequently his laser had top leads that were *short* as he saw it but still too long for success. The laser would not work. . . . Eventually, he found that in order to make the leads short enough to count as *short* in 'TEA laser society' he had to go to a great deal of trouble. The large and heavy top capacitors had to be mounted above the bench in an inverted position in a strong and complex steel frame so that their terminals could be very near to the top electrode. . .

Later, as this aspect of the design of the laser came to be understood as a piece of *electronics* it became quantitatively theorized. The *inductance* of these top leads was then seen as the crucial variable. It then became accepted and that these top leads had to be 8 inches (20cm) or less in length." (Collins, *Artificial Experts*[1990], 111–13)

Could All Relational Tacit Knowledge Be Made Explicit?

Relational tacit knowledge can be told if only the contingencies would go away. But can the contingencies be made to go away? Concealed knowledge could be entirely eliminated if everyone stopped keeping secrets, running secret societies, and exploiting apprentices. But these changes are no more likely than the elimination of crime. In fact, it is somewhat depressing that even the sectors of society driven by high moral virtue, such as science, still have their secrets. Something similar applies to logistically demanding knowledge—it will always be too expensive to make the whole of it explicit.

Can all knowledge that that has to be transferred through personal contact as a result of mismatched saliences be made explicit? Only if we could make everyone privy to the contents of everyone else's heads, but again, this is logistically impossible outside of very small groups who live or work together for much of their lives, such as husbands and wives, military platoons, and sports teams.

What about unrecognized knowledge? To recognize all currently unrecognized knowledge would involve completing the task of science and understanding every mechanism that controls the material world. Again, while there is no principled reason that any particular piece of unrecognized knowledge could become recognized, one cannot imagine *every* piece becoming recognized. There will always be a frontier with things just beyond it waiting to be discovered and told.

Thus, the way human societies work plays a role in our understanding of the irreducibility of the tacit. In society as we know it there will always be secrets, mismatched saliences, and things that are unknown but may be about to become known. This fact has to do with the way society organizes itself rather than having anything to do with the intrinsic nature of the social (the intrinsic nature of the social will be discussed in chapter 6). Nothing of what we have discussed so far prevents any specific piece of tacit knowledge from being told and this makes it still more clear that principles to do with the nature of knowledge are not at stake. The prohibition is only in respect of all the relational tacit knowledge being explicit at one time. Relational tacit knowledge is tacit because of the contingencies of human relationships, history, tradition and logistics. This, however, gives us another reason for referring to it in everyday speech as tacit: the fact is that whatever you do there will always be knowledge that is not made explicit for these contingent reasons and it, therefore, will be an ever-present feature of the domain of knowledge as it is encountered even though its content is continually changing. To summarize, any one piece of relational tacit knowledge can be made explicit, because the reason it is not explicit is contingent on things that can be changed. But all relational tacit knowledge cannot (*logistic practice or higher*) be made explicit at once.

FIVE

Somatic Tacit Knowledge

> The ineffable domain of skilful knowing is continuous in its inarticulateness with the knowledge possessed by animals and infants. . . . The anatomist exploring by dissection a complex topography is in fact using his intelligence very much like a rat running a maze. . . . We may say in general that by acquiring a skill, whether muscular or intellectual, we achieve an understanding which we cannot put into words and which is continuous with the inarticulate faculties of animals. (Michael Polanyi, *Personal Knowledge,* 90)

Bicycle Balancing

Because Polanyi used it as a central example, bicycle riding has become a paradigm of tacit knowledge. When we ride our bikes we do not self-consciously use any physical or mechanical models; somehow, with practice and training, the ability to balance on a bike becomes established in our neural pathways and muscles in ways that we cannot speak about. We do not learn bicycle riding just from being told about it (coaching rules and second-order rules aside), or reading about it, but from demonstration, guided instruction, and personal contact with others who can ride—the modes of teaching associated with tacit knowledge. That is why we say our knowledge is tacit—we cannot "tell it" but we can have it passed on in ways which involve close contact with those who already have it.[1]

1. We could, of course, buy a bike, read about how bike riding is done (we might read second-order descriptions or coaching hints—see chapter 3), and then rediscover how to do it ourselves. Rediscovery of nonsocial kinds of tacit knowledge is always a possibility because someone had to discover it in the first place so it can happen again. But rediscovery, even when it is aided by written hints, is not transmission of knowledge.

Notice that balancing on a bike is learned through a process that might well be referred to as "socialization." One picks it up through immersion in the society of other bicycle riders, although there is often some more directed teaching going on too. There is a phrase concerning certain kinds of skills: one says that it is "just like riding a bike." What this means is that it is a skill that once learned is never forgotten—the body's "knowledge" of this skill is never lost. Presumably, what happens in the case of these abilities is that some part of the brain and some part of the related nerve pathways and muscles are changed in a fairly permanent way. These parts of the brain and related bits of physique change in the same way that a weightlifter's body changes as the muscles grow in the course of training. It is a matter of condition 4 transfer. The fact that it is condition 4, rather than conditions 1, 2, or 3, is one reason for calling the knowledge tacit; on its own, no amount of string transformation or enhancement will transfer the abilities. Of course, one can lose the knowledge because those new parts of the physical being can deteriorate, but the deterioration would be like the deterioration of the weightlifter's muscles, not like the change in the signals coursing through a computer's circuits when Microsoft XP is exchanged for Linux or like the changes that occur as a result of a native language speaker's loss of contact with the flux of language in society (see below).

Two of Polanyi's comments about bicycle riding are quoted in box 11. The second comment in this box reveals that something called "the rules of bike riding," in the sense of bike riding as discussed by Polanyi—which I am going to call "bike balancing"—can be told in sense 4 of explication. Polanyi explicates them himself just three pages after he says they are tacit. Furthermore, these rules, or some set of rules, could, in principle, be programmed into a mechanical bike balancer with elaborate feedback circuits—so they are also explicable in sense 3.[2]

I also claim that if our brains and any other elements of our physiology involved in balancing on a bike worked a million or so times faster, or, what is the equivalent, if we rode our bikes on the surface of a small asteroid with almost zero gravity so everything happened much slower, we ourselves could probably use Polanyi's rules to balance. Under these circumstances, balancing on a bike would be like assembling flat-pack furniture: as we began to fall to the left or the right we would consult a booklet and slowly adjust the angle of steering according to the instructions for remaining upright. Under these circumstances, we would have come to ride

2. Mechanical bike balancers have been built but so far as I know they use gyroscopes in ways not covered by Polanyi's rules. The principle stands, however.

Box 11. Polanyi on bicycle riding.

"If I know how to ride a bicycle . . . this does not mean that I can tell how I manage to keep my balance on a bicycle. . . . I may not have the slightest idea of how I do this, or even an entirely wrong or grossly imperfect idea of it, and yet go on cycling . . . merrily. Nor can it be said that I know how to bicycle . . . and yet do *not* know how to coordinate the complex pattern of muscular acts by which I do my cycling. . . . I both know how to carry out [this performance] as a whole and also know how to carry out the elementary acts which constitute [it], although I cannot tell what these acts are."

"In order to compensate for a given angle of imbalance α we must take a curve on the side of the imbalance, of which the radius (r) should be proportionate to the square of the velocity (v) over the imbalance $r \sim v2/\alpha$." (Michael Polanyi, "The Logic of Tacit Inference," *Philosophy* 41, no. 1 [1966]: 1–18 [quotes, 4, 6–7])

by condition 3 transfer—the use of a long string. (Balancing on a bike is a mimeomorphic action!)

In sum, in the normal way, what stops bike balancing from being carried out according to explicit instructions is the limit on the speed of our brains and our reactions. It is the limit imposed by our physiology. For this reason we normally have to learn bike balancing through the means typically used to transmit tacit knowledge. Bike riding depends, then, not on unadorned tacit knowledge, but on what we call somatic-limit tacit knowledge. This is knowledge that is tacit because of our bodily limits even though it can be explicated in the third and fourth senses of explication.[3]

The Dreyfusian Five-Stage Model

One thing we already know is that, irrespective of whether humans can use sets of instructions to ride bicycles or whatever, they do many things better if they do not process the instructions with the conscious mind. The person most well-known for championing this view is Hubert Dreyfus. With

3. Mackenzie's (2001) discussion of the conflict between computer proof and human proof in mathematics is another illustration of somatic-limit tacit knowledge. If human brains had greater capacity there would be no conflict because the brain would be able to grasp the computer proof. Note, the proof remains a proof whether done by computer or brain; one sees how misleading the concentration on human abilities can be when it comes to trying to understand knowledge rather than humans.

Box 12. The Dreyfusian five-stage model of skill acquisition.

Stage 1: The *novice* driver will try to follow explicit rules and as a result the performance will be labored, jerky, and unresponsive to changes in context. The skill will be exercised "mechanically," following rules such as "change gear when the car reaches 20 mph as indicated on the speedometer."

Stage 2: The *advanced beginner* masters more unexplicated features of the situation, such as using the sound of the engine as an indicator of when to change gear.

Stage 3: *Competence* is achieved as the number of "recognizable context-free and situational elements" becomes overwhelming, and expertise becomes much more intuitive rather than calculating.

Stage 4: The *proficient* driver recognizes whole traffic scenarios "holistically" in the same way as the advanced beginner recognizes, say, engine sounds.

Stage 5: *Expert* status is achieved when complete contexts are unselfconsciously recognized and performance is related to them in a fluid way using cues that it is impossible to articulate. Hence the common experience of driving a familiar journey to work and being unable to remember anything about it after one arrives.

(Hubert L. Dreyfus and Stuart E. Dreyfus, *Mind Over Machine* [1986], 16–36)

his brother Stuart, he worked out a five-stage model of skill acquisition that turns on progress from the use of explicit rules to the internalization of the physical skill. A typical example invoked by the Dreyfuses is learning to drive a car: the driver initially controls the car according to instructions about when to change gear and so forth and progresses through a series of stages to grasp entire situations without conscious thought, even driving familiar routes without being aware of the journey (see box 12).

The first problem with the schema is that though the stages are treated as a general description of all human skill acquisition they do not apply to all skills. For example, if it was bicycle riding that was under discussion the first two stages would not apply. In bicycle riding there is, as it were, a protostage—stage 0—when one falls off, and everything else is steady improvement from stage 3 onward. At no time are explicit instructions pertaining to the maintenance of balance (coaching and second-order rules aside) of any use to a human learning to ride a bicycle (outside of low-gravity conditions). In the case of copy typing, on the other hand, there is nothing much after stage 3 beyond the social sensibilities required to live in society applied to typing—not something that one would expect a machine to be able to mimic (see chapter 6) but not something that is learned during the process of learning typing, per se.

The most interesting thing about the model is the contrast between conscious and unconscious processing. We can agree that humans can drive cars and do a subset of other skills in two different ways: one way, mostly useful only when learning, is through self-conscious attention to explicit rules; the other way, that of the expert in the case of most skills, is without paying much or any self-conscious attention to the process. As the Dreyfuses say, the second way is usually more efficient for humans than the first way.

Before analyzing the point further let us exemplify it with a striking case of its consequences for skilled touch typists. If you, reader, are a skilled touch typist—that is, someone who does not look at the keyboard as they type—take a scrap of paper and, without looking at the keyboard, try to write out the letters of the alphabet in three rows corresponding to the pattern that is found on the keyboard. The chances are you have no idea where the letters are—I certainly didn't when I tried it. Yet my fingers, or my fingers along with my brain and whatever components of my body are involved, must "know" where all the letters are, because I am typing this passage fairly accurately without any awareness of what is happening at the level of the keyboard.[4]

A related finding is that skilled copy typists seem to be able to type as fast and as accurately—or nearly as fast and as accurately—whether they are typing sense or nonsense. I, on the other hand, am a fairly accurate touch typist so long as I am typing out my own thoughts, but not a very good copy typist of other peoples' words, and I find it almost impossible to copy-type nonsense. In other words, when I type, meaningful words have to

4. This is my version of a widely known but unpublished experiment done by Jonathan Grudin in which he asked typists to correctly replace keys which had been removed from a typewriter keyboard. Thanks to Grudin for endorsing my simplified version of his idea (personal communication, December 26, 2007).

"pass through" my head if I am to be efficient, whereas skilled copy typists can manage without a meaning stage, or much of a meaning stage.[5]

All this goes to show that one cannot treat skills learning as monolithic after the Dreyfusian fashion. There different ways of accomplishing what, to an outside observer, might look like the same skill and there are, to repeat, lots of different stages involved in learning different skills.

What remains important is that skills cannot generally be executed with the same efficiency by humans if they are paying self-conscious attention to the rules through which they were taught, but once more, this seems to bear on nothing but the way humans work; it does not bear on the way knowledge works. Here the typing example is more philosophically pertinent than the overcomplicated car-driving example. Granted, we humans cannot generally type as fast or as efficiently when we are paying attention to the keys but that's just us. An automated typing machine that scanned print that was set out in a clear and undamaged font, transformed it into editable text, and then typed it out again could work as fast and as accurately as any human typist, and faster if desired (which is, of course, what modern scanners do but without the intervention of the keyboard). The constraints on the methods available for efficient typing by humans are somatic limits; they have everything to do with us and nothing to do with the task as a task—nothing to do with the knowledge as knowledge.

A serious source of misunderstanding of tacit knowledge, including the analysis of the limits of artificial intelligence for which Hubert Dreyfus is justly famous, is an obsession with the human body at the expense of the proper obsession, which should be with the nature of knowledge. That part of tacit knowledge, which is tacit because of the nature of the human body, is just one part of the domain of tacit knowledge and a part that is not particularly hard to understand. It is also something that is generally easy to automate. The limits of the human—somatic-*limit* tacit knowledge—have little or no bearing on the possibilities of mechanization or artificial intelligence. In all the ways that do not involve the way we intentionally choose to do certain acts and not others, and the way we choose to carry out those act, the human, per individual body and brain—that is, the human per complicated animal (that is, the human engaged in mimeomorphic actions)—is continuous with the animal and physical world. We are just like complicated cats, dogs, trees, and sieves. When doing these things we are just

5. I tried some experiments along those lines and obtained near-perfect results first time around but never managed to get things to work quite so well again—hence the qualification of "much of a meaning stage."

complicated sets of mechanisms (which become mysterious only if we start to try to describe our experiences—to make them explicit).

Sometimes we can do things better than cats, dogs, trees, and sieves can do them, and sometimes worse. A sieve is generally better at sorting stones than a human (as a fridge is better at chilling water), a tree is certainly better at growing leaves, dogs are better at being affected by strings of smells, and cats are better at hunting small animals. We are better at calculating and, perhaps, copy-typing, though there is no reason of principle why a pigeon could not be taught to copy-type by looking at print and pecking at the corresponding keys for a reward. It is all a matter of one kind of machine versus another and the most important thing that stops us seeing this is our propensity to try to describe what we do and pass it on to others through the transmission of strings. To repeat a theme found in the introduction, if we were to stop talking and just get on with things—that is, if the tacit was not made mysterious by its tension with the explicit—there would be no puzzle at all about the body, per se. That teaching humans to accomplish even mimeomorphic actions is a complicated business, involving personal contact, says nothing about the nature of the knowledge, per se. The mistake is to see all problems of human knowledge acquisition as problems of knowledge.

Incidentally, some of the things that machines (if the right machines existed) could do better are sometimes done by human beings as a test of their mimeomorphic abilities. An example is fast repetition of the following:

> I'm not a pheasant plucker,
> I'm a pheasant plucker's son,
> And I'm only plucking pheasant,
> Till the pheasant pluckers come.

This is an especially interesting case because it is one where humans can do the task best, or only, if they do pay self-conscious attention to the process of uttering the words, violating the rule that we execute all skills better if we do not attend to what we are doing. The same goes for remarkable feats of memory or spelling difficult words, and it probably goes for certain physical skills, such as competitive high-board diving or synchronized swimming, where the idea is to act as though one is a machine. I would guess that self-conscious proprioception is likely to be central to success in these cases. Something similar probably applies to many "dead ball" games, such as golf, where the performer often has a coach (in golf it is the coach off the course and the caddy on the course) to share the self-conscious calculative activities that are an undoubted part of the activity.[6] But, if we strip out the social aspects of the games (of which more later) nearly all sports, even many of those we play best without self-conscious attention, could, in principle, be executed by machines. That, in practice, the prospect is remote should not cut any philosophical ice except for those who are more interested in the human organism than in the nature of knowledge.

Returning to the question that, as we remarked, "haunts" any treatment of tacit knowledge, we want to know whether all human abilities are in principle describable in terms of sets of rules that could be executed self-consciously by humans, perhaps in slowed-down circumstances—or by different kinds of devices such as computer controlled machines. Polanyi has given us the rules for balancing on a bicycle, but no one has given us the rules for balancing on, say, a unicycle. Nevertheless, there is every reason to think that unicycle balancing rules, like bicycle balancing rules, are explicable knowledge (in senses 3 and 4), though, again, the set of explicable rules would not be usable by any human except in very special "low-gravity" circumstances. Are all human abilities like this or are there things that bodies and brains can do that could not be described, even in principle? Chess is one example where such a thing has been claimed for human abilities.

Chess

Chess is one of the key points of discussion for the notion of artificial intelligence. Hubert Dreyfus famously claimed that no computer would ever beat a chess grand master because it was impossible to set out and program the rules used by grand masters. He was wrong in that a program running on a computer that came to be known as Deep Blue, developed by IBM, did beat grand masters, and we now understand that a powerful enough computer appropriately programmed is likely to beat the best human chess player most of the time. But there is another sense in which Dreyfus was not wrong. As in the case of bicycle riding, one must attend carefully to the way the human on the one hand, and the computer on the other, carry out the task. The point is made in box 13:

To cut a long story short, the key idea of a computer game-playing algorithm is "brute strength" calculation of the outcome the available alter-

6. As a bad golfer myself, I often complain when I come off the course that in spite of my being a university professor it is my stupidity at not choosing the right shots to play that has been as much a cause of my poor score as my inability to play the shots I choose.

Box 13. Dreyfus and chess.

In 1972, Hubert Dreyfus wrote: "In chess programs, for example, it is beginning to be clear that adding more and more specific bits of chess knowledge to plausible move generators, finally bogs down in too many ad hoc subroutines. . . . What is needed is something which corresponds to the master's way of seeing the board as having promising and threatening areas." (Quoted in Hubert Dreyfus, *What Computers Still Can't Do* [1992], 296)

In 1997, Dreyfus wrote: "I said that a chess master looks at only a few hundred plausible moves at most, and that the AI people would not be able to make a program, as they were trying to do in the '60s and '70s, that played chess by simulating this ability. I still hold that nothing I wrote or said on the subject of chess was wrong. The question of massive brute-force calculation as a way of making game playing programs was not part of the discussion, and heuristic programs without brute force did seem to need, and still seem to need, more than explicit facts and rules to make them play better than amateur chess. But I grant you that, given my views, I had no right to talk of necessity." (From "Artificial Intelligence," a debate between Daniel Dennet and Hubert Dreyfus, *Slate* magazine, May 1997, http://slate.com/id/3650/entry/23905/)

native future moves. If all *possible* combinations of moves by both players are known from the outset then whoever makes the first move can win (or draw) the game. Simple games like noughts and crosses (tic-tac-toe) can be completely evaluated in this way and a computer can always win or draw if it makes the first move (indeed, tic-tac-toe is simple enough for even humans to win in this calculative way). The game of drafts, or checkers, after many decades of work, has also just been made to yield to such an exhaustive program.[7] Chess, however, is so complicated that it is logistically impossible to work out all possible combinations of moves in advance—such a calculation would need a computer that would be vast compared to the universe. Grand masters do calculate a few moves ahead but mostly they win by seeing patterns of play that they cannot fully explain. It was because Dreyfus understood that the chess problem was not practically solvable by exhaustive brute force methods, and that no one knew how to write a program that would recognize patterns in the human way, that he said

7. Cho 2007.

no computer would ever beat a grand master. On the other hand, big computers, being better at calculating than humans, can see a few more moves ahead than grand masters even if they cannot see all the way to the end of the game. Importantly, the calculation problem is so huge that even to see a few more moves ahead than a human the big computers have to restrict the search tree by imposing a few humanlike "heuristics" or "rules of thumb" that indicate whether one position is better than another in some general sense such as keeping command of the center of the board. What surprised everyone was that being able to calculate only a few extra moves ahead was good enough to win. In terms of being able to win, chess, it turned out, was not quite as intractable a game as Dreyfus and most other people thought it was.

The artificial intelligence community celebrated their victory over their archenemy (at that time) Dreyfus.[8] But, if the object of the artificial intelligence exercise is to model the human mind, then the chess victory was hollow—chess computers do not achieve victory by mimicking the way humans play. Furthermore, the different styles of play—computer versus human—are recognizable on the board. Only if the object was to win at chess by any legitimate means were Dreyfus's arguments, as opposed to his (perhaps misjudged) challenge, defeated.

If we define good chess playing simply as "the ability to win," then chess playing is another example of somatic-limit tacit knowledge. Humans play chess the way they do because that is the only way they can do it. To see the similarity with bike balancing, we need only carry out the equivalent of the bicycling thought experiment for chess: let the game take place in the equivalent of low-gravity conditions by removing the chess clock and extending the lifetime of the human players. Each player could then take as long as they liked for each move—for example, a hundred years—and this would allow them to evaluate the outcomes of possible future moves as far ahead as can today's computers. If today's grand masters could calculate this far ahead that is the way they would play, because it would enable them to win more games.[9]

But what about the way humans actually play chess? Is it imaginable that the methods that humans actually use for playing high-level chess could be transformed into a set of explicit instructions, or is there some-

8. After many decades Dreyfus's ideas are now been absorbed into mainstream of artificial intelligence thinking.

9. Whether chess would be such a popular and "iconic" game if it was merely a test of calculative ability is less certain.

thing special about the human organism? Is it that human intuition, as Dreyfus terms it, is just not explicable? Is it that there is not only something called somatic-limit tacit knowledge, which turns on the limitations of the body, but something called somatic-affordance tacit knowledge? Somatic-affordance tacit knowledge would turn on the special physical nature of the body (and brain).

Insofar as we are concerned with the body and brain as material substances it is possible to approach an answer by starting with sieves. Let us say that we have a pile of randomly shaped stones and we put them on a certain sieve, with a certain number of circular holes of a certain diameter, and shake it for a certain length of time. We know from experience that the pile of stones will be separated into two groups. Another way to separate the stones into two groups might be to measure the stones and sort them according to their minimum diameter, for only stones with a minimum diameter less that the diameter of the holes would go through the sieve. It turns out, however, that the second method would reproduce what the sieve does only if (a) the stones were spherical and (b) the sieve was shaken for a "pretty long time." But if the stones were randomly shaped then the result of sieve sorting is a complicated function of many things, including the distribution of the shape of the stones and the time of shaking. Consider that the stones were long and thin, like spaghetti. One might shake for a very long time before *any* of the smaller stones fell through, even though measurement would show that they could pass through the holes. So what would one measure, and how many measurements would one need to take and how one would relate these measurements to the length of shaking to get a result based on calculation that was the same or very similar to that produced by the sieve? And we still have not taken into account the degree of vigor of the shaking, the detailed surface topology of the stones, their coefficients of friction with each other and with the material of the sieve, and the number of stones in relationship to the number of holes and their distribution on the sieve, all of which would affect the likelihood of any one stone appearing in the right orientation just above a hole and slipping through in during the time of shaking. And in spite of the difficulty of the logistics, the sieve is a really easy case of these things. So the sieve case is not so different from the chess case—there is a way of reproducing what the sieve does but it is very hard to do it in just the same way as the sieve would do it. That is why we tend to use tools to do many jobs rather than programmed machines. As argued in chapter 1, in principle it could all be done digitally but in practice it is easier to use a device using an analogue string so long as it is producing the result we want.

In the case of something as complicated as the human brain and sensory mechanisms it might well be that we cannot (*logistic practice*) do some of the things it does by using other substances and mechanisms and we might never be able to do it. But though the lived experience of doing such tasks is reasonably described by terms such as "intuition," there is no more in the way of reasons of principle to think that these tasks are any different in kind to those done by cats, dogs, trees, and sieves than there are reasons of principle to think that a rocket traveling at 280,000 kilometers per second is different in kind to one traveling at 10 kilometers per second. There are things out there in space traveling at 280,000 kilometers per second, but we cannot make such things using rocket technology, and there are brains seeing patterns in a game of chess, but we cannot make such things using silicon technology, but otherwise there is nothing special going on. There is nothing any more special than would be involved in, say, trying to make a moon rocket out of ice—one knows it could be done in principle but the practicalities would be almost beyond contemplation. Once more, the argument is confounded by mixing up the analysis of the way humans do things with the nature of the knowledge itself.[10]

In sum, there is nothing that has been said so far that would prove that the human way of chess playing could not, in principle, be described in terms of a mechanical sequence instantiated in the particular substances of brain and body. There is nothing to prove it could not be explicated in senses 3 and 4. Even something so seemingly irreducible to so-called intuition as spotting the analogy between the broad position in one game and the different in detail but broadly similar position in another can be thought of mechanically—imagine the brain has mechanism that works a bit like a set of templates so that there is an area of tolerance around what will fall through a particular "analogy template." When two slightly different items can fall through the same complexly shaped "hole," we call it an analogy. It just happens, however, that we don't know how to construct these templates artificially—there they are in the human brain but we don't know how to make them. Perhaps the brain works like some kinds of ana-

10. Evan Selinger claims that emotions are central to human abilities, such as bringing up children (personal communication, December 2007). It seems likely that he is right and it could be that the ability to have emotions is to do with somatic affordance. (Note that this particular aspect of somatic affordance seems to stretch to animal bodies too—the fight-or-flight reaction and so forth.)

Box 14. The irreducibility of the analogue?

There may be solutions to problems that can be found by analogue processes but not by digital (or there may not). For example, *n* towns have to be connected by roads whose total length is a minimum—roads may meet at "vertices" that are not in towns if this leads to an overall shorter road length. There is no algorithm known for solving this problem that takes less than 2^n steps, which means that as soon as the number of towns becomes large the problem becomes intractable (there are around 2^{125} particles in the universe). But, as A. K. Dewdney explains: "Attach two parallel sheets of rigid transparent plastic to a handle and insert pins between the sheets to represent the [towns] to be spanned. Now dip the gadget into a soap solution and withdraw it. There before your eyes is a soap film connecting the pins [with extra vertices as necessary]." Unfortunately, though this will represent the minimum length of soap film for the number of towns and extra vertices that have been created, it is not certain that this is the optimum solution: a still shorter solution might perhaps be found with a different number of vertices—it is impossible to be sure. (A. K. Dewdney, "On the Spaghetti Computer and Other Analog Gadgets for Problem Solving" [1984], 18–19)

logue computer, the processes of which are either very hard to reproduce in digital computers or with different substances, as is argued in box 14.

Something similar may apply to the human body. Even if the physics, chemistry, and mechanics of the body were fully understood, just as in the case of the brain, it might be that a device that reproduces what it does could not be made from substances other than those from which the naturally occurring body is made. It might just be that the elasticity, shock absorbency, and degree of stick of the tendons is just right for controlling the limbs in certain circumstances and it will turn out that reproducing this though calculation of accelerations and decelerations and implementing them in objects made from different substances is an intractable problem. The philosophical point is the same whether it is brain or body that is under consideration.

For the last eighty years or so, a great deal of philosophical effort has gone into understanding the relationship between the human body and the material world it inhabits. For example, Martin Heidegger has written about the "readiness to hand" of tools like a hammer—something that is used with no more consciousness than the tendons in the arm that wields it unless it is brought to the attention by being broken and Maurice Merleau-Ponty has talked about the way a blind person's stick becomes part of the blind person's body.[11] In more recent years the artifacts of thinking, such as the mathematician's paper and pencil, have been treated by, say, Andy Clark, as extensions of the brain—Clark says that thinking leaks out beyond the skin and skull into the scaffolding of the material world. And, as we have seen, the argument works in reverse and allows us to treat the brain as a tool. Take the case of the instructions for solving differential equations transmitted to a poor mathematician. Surely they would work only if much more in the way of brain pathways, chemicals, and even substance were transmitted too; that is where mathematical ability is found. It looks, then, as though tools, body substance, and brain substance can all be thought of as belonging in the same "logical space" when it comes to knowledge transfer.[12]

In sum, we can visualize the body and the tools or material scaffolding within which the brain and body work, and even the brain itself, as components belonging to an analogue string-driven device. Because analogue string-driven devices have special properties that turn on the materials of which they are made and the shape in which they are formed, it might be very difficult to reproduce the outcomes they produce by other means for logistical if not for logical reasons. This remains the case even though reproduction of outcomes can be produced without full understanding of processes, as in the case of the record-and-playback paint sprayer and similar devices such as neural nets.

Put the special nature of the body as material together with the special nature of the brain as material and one understands how hard it will be to make a machine to play soccer even if one forgets about the social relations with the other twenty-one people on the field of play.

But maybe one day it will be done, or at least understood. There are all kinds of things we cannot make using alternative materials and methods. We have noted that is rather difficult, if not impossible, to reproduce something as simple as a sieve in any other way; it is still more difficult to reproduce a tree and more difficult still to reproduce a cat. Humans-as-animals

11. My comments on Heidegger are taken from my understanding of Dreyfus's interpretation of his philosophy; Dreyfus's interpretation has been criticized, but I am unable to comment on this debate.

12. I am grateful to Andy Clark (Clark 2003; Collins, Clark, and Shrager 2008) for leading me to this point. Along the way, he refers to insights from David Chalmers about the relationship between the self and the brain (Collins, Clark, and Shrager 2008).

are just one further step along the way. And that realization may lie behind the confidence that the artificial intelligence community has in their ability to understand us. Furthermore, the theory of evolution provides a conceptual scaffolding for this kind of ambition—if animals, including humans, just evolved as a result of the interplay of chance and natural causal sequences, then why should we not be able to reproduce, or at least understand, the outcome? Or so the argument goes. Indeed, techniques like neural nets and evolutionary programs seem to reproduce the very mechanisms as well as the end products.

To sum up, the two major subdivisions of somatic tacit knowledge are somatic-limit tacit knowledge and somatic-affordance tacit knowledge. They pull in opposite directions. Somatic-limit tacit knowledge is what prevents humans coming to be able to execute certain actions or acquire a certain competence as a result of the transmission of a string even though that string is an explication of the knowledge that corresponds to that action or competence. In other words, somatic-limit tacit knowledge prevents knowledge that is explicable in sense 4 (see table 4) from becoming explicable in sense 1, because the substances of body and brain are not capable of being affected by that string in the way that, say, a computer, with a processor much faster than a human brain, could be affected by it (it is explicable in sense 3). Somatic-affordance tacit knowledge prevents (in the current state of the art) a string which counts as an explication in sense 4 affecting a device *other than* a human, because only the human body and brain are made of suitable substances. In other words it prevents sense 4 of explicable from becoming sense 3 of explicable.

The Asymmetry of Self and Material Scaffolding

On the basis of this argument it seems as though the dualism of brain and body and the dualism of body and its material "scaffolding" must both be abandoned. And yet there remains an echo of the old Cartesian dualism between *mind* and material. The relationship between the "self" and its tools, including the brain and the body, is not a symmetrical one. "I" can repair or replace any of the bits of the ensemble. I can replace my hammer with a heavy stone. If the handle of my screwdriver is wrapped in barbed wire I can remove it or wrap it in a cloth. If I lose my arm I can replace it with a prosthesis. If my muscles are too weak to lift heavy weights I can use a jack or an exoskeleton. If I lose the power of speech I can, like Stephen Hawking, use a voice generator. The phrase used by Merleau-Ponty to describe the blind man's relationship to his stick (box 15) is simply wrong.

Box 15. Merleau-Ponty and the blind man's stick as an extension of the body.

"The blind man's stick has ceased to be an object for him, and is no longer perceived for itself; its point has become an area of sensitivity, extending the scope and radius of touch, and providing a parallel to sight. . . . To get used to a hat, a car, or a stick is to be transplanted into them, or conversely, to incorporate them into the bulk of our own body. Habit expresses our power of dilating our being in the world, or changing our existence by appropriating fresh instruments." (M. Merleau-Ponty, *Phenomenology of Perception* [London: Routledge, 1962], 143)

The stick may extend "the scope and radius of touch" but it is not true that "to get used to a hat, a car, or a stick is to be transplanted into them." We are not transplanted into them—they only are transplanted into us. The relationship is always asymmetrical. By this I do not mean to dispute Merleau-Ponty's phenomenological description of the relationship between the blind man and the stick, merely to point out that we only ever describe it from the point of view of the man, not the stick. We do not say anything like, "the blind man has ceased to be an object for the stick, he has now become transplanted into it, extending the stick's area of sensitivity to engagement with the full richness of the human world." In other words, we can use the stick as an extension but the stick cannot use us.[13]

And there is no reason to think that sometime in the future, if not now, if bits of my brain are deficient, I will not be able to replace them with something artificial. Even more remarkable, as has been established, the self can choose to use the brain in different ways, sometimes its internal processes being unconscious and sometimes carrying out what is roughly the same task through more calculative processes—presumably these different modes of working use different components within the brain. So bits of the brain can be substituted one for another. The hammer, screwdriver, computer, and so forth can do none of this if there is something wrong with "me," so there is still a role for the self.

13. The so-called actor-network theory has succeeded brilliantly in the academic marketplace by cleverly failing to acknowledge this obvious asymmetry and claiming that its absence from the theory represents a philosophical insight. For critiques of actor-network theory, particularly in respect of its failure to deal with the social nature of human existence or the special element of human knowledge which is tacit, see Collins and Yearley 1992 and Collins 1998. For another critique, see Bloor 1999.

Box 16. Badly broken text that is and isn't easily repaired.

blveiee hmuan I aulaclty waht rdanieg I. The cdnuolt mnid, phaonmneal of the i to a aoccdrnig Cmabrigde dseno't it oerdr waht the taht ltteres in a is are, the wrod iproamtnt taht frsit the and ltteer pclae be in the cluod rghit. The can rset rscheearch olny at be Uinervtisy, a taotl mses lsat and you tihng mtaetr in can sitll pweor it whotuit a pboerlm. Tihs not is raed uesdnatnrd the bcuseae deos mnid raed by was lteter istlef, but the as a wrod Azanmig wlohe.was yaeh huh?and I awlyas tghuhot slpeling ervey ipmorantt!

I cdnuolt blveiee taht I cluod aulaclty uesdnatnrd waht I was rdanieg. The phaonmneal pweor of the hmuan mnid, aoccdrnig to a rscheearch at Cmabrigde Uinervtisy, it dseno't mtaetr in waht oerdr the ltteres in a wrod are, the olny iproamtnt tihng is taht the frsit and lsat ltteer be in the rghit pclae. The rset can be a taotl mses and you can sitll raed it whotuit a pboerlm. Tihs is bcuseae the huamn mnid deos not raed ervey lteter by istlef, but the wrod as a wlohe. Azanmig huh? yaeh and I awlyas tghuhot slpeling was ipmorantt!

(Thanks to Terry Threadgold for forwarding the second paragraph.)

But what is this self? Let us take another look at what it does. One way in which the asymmetry of the brain and its tools is exhibited is in what I will call the ability to "repair." It has already been referred to where examples such as the speech transcriber have been discussed. Listen to ordinary speech and it is enormously noisy and broken, yet humans have the ability to make a pretty good job or repairing what they hear by reference to its sense. Though automated machines can also repair broken speech and render it into type, the mistakes they make are very different in kind to those made by humans. The same capacity can be illustrated if written text is sufficiently badly broken to need repair. Box 16 shows a particularly dramatic example of our everyday ability to make good something broken while hardly thinking about it so long as it makes sense to us. Most readers will find it very hard to read the first passage in box 16, if they can do it at all, but that they will be able to read the second paragraph with almost as much ease as they could read unbroken text.

The "words" are exactly the same in both paragraphs, but in the first their order has been rearranged so that the passage does not make sense.

There is not one, but three remarkable things about this demonstration. The first remarkable thing is that most readers can read the second passage so easily and the overall demonstration shows that the reading is accomplished via meaning. The second, still more remarkable thing is that, though my spell-checker highlighted almost every word with a jagged red line indicating a mistake, the copy editor of this book will not even think about correcting either paragraph. That is an effect of understanding meaning at a still higher level—the meaning of a whole group of paragraphs.

The third remarkable thing is that you, dear reader, have actually undertaken to try to repair both passages, known when it was time to give up in the case of the first passage, and known that you should persevere in the case of the second passage in spite of any initial difficulty.

The task of any really deep artificial intelligence project must be to reproduce all of this, and the task of any attempt to make tacit knowledge explicit must be to understand how it is done.[14]

The asymmetry between me and my tools lies in the fact that I can do this kind of thing in respect of the things around me, but the things around me cannot do this kind of thing in respect of me. Here is the echo of the Cartesian self. That it often goes unnoticed by careless observers is precisely because of our exquisite repair abilities. Hardly noticing the differences between our speech transcribers, spell-checkers, dogs, and ourselves, we repair the mistakes they make and count them as like ourselves. They become, in the term introduced earlier, "social prostheses": entities that can take the place of a social being because the rest of the social organism fills in the gaps. The point was made in box 6 (p. 71). There it was explained that it was tempting to think that pocket calculators "do arithmetic." But they do not. We think they do only because we engage in so much unconscious repair.

What is the self? The question is hard but it is possible to answer an easier question. At least one major component of the self (and it may be the entirety of the self), is society. My ability to repair the broken text (along with my ability to approximate) depends on understanding what is proper English and what is proper approximation in the appropriate contexts. The fact that the reader has persevered or not persevered, to just the right extent for the sake of the demonstration, to accomplish the repair of the English, and the fact that the editor has *not* done any repair work on the English, is to do with their ability to use the language with acute sensitivity to

14. And any theory that treats the human and nonhuman in a symmetrical way is in thrall to the solving of the problem.

context. It is a matter of condition 5. These abilities are learned from embedding in society. In this case it is embedding in the world of the English language and the world of academic argument. Another way to say this is that humans and humans alone have the ability to execute polimorphic actions successfully.

Take away computers and social life would be different. My very writing of this book would feel very different if I could not make use of this laptop computer with its grammar-checker and its spell-checker, the facility it gives me to inspect my past works, and the works of others through the Internet connection, and the ability to cut-and-paste materials from many sources. And yet though these things may *contribute* to my social experience they cannot *absorb* anything equivalent to my social experience. I "love" my computer and its spell-checker and find them an enormous help to my life, but they do not participate in the world of language in the way that I participate; they participate only in the world of string transformation. The next chapter takes up these themes.

Could All Somatic Tacit Knowledge Be Made Explicit?

We ask the standard question once more: can all somatic tacit knowledge be made explicit? The answer is "yes," if here "explicit" means "expressed scientific understanding of causal sequences." Somatic tacit knowledge is just causal sequences and in principle, if not in practice, these can be understood scientifically. We can foresee how we might go about it though it remains technically beyond our capacity. The answer is also a qualified "yes" if we are talking of explicable knowledge (in sense 3)—the ability to reproduce the uses of the knowledge in machines. The limits are not limits of principle or ability to foresee how to do it; they are limits of practical ability, the affordance of different materials, and engineering considerations of that kind. In sum, there is nothing philosophically profound about Somatic tacit knowledge, and its appearance of mystery is present only because of the tension of the tacit with the explicit: if we did not feel pulled toward trying to say what we do, and if we did not make the mistake of thinking this is central to the understanding of knowledge, we would find nothing strange about our brains' and bodies' abilities to do the things we call tacit. And that is why too much concentration on the body as the seat of the tacit takes one away from a proper understanding of the idea.

SIX

Collective Tacit Knowledge and Social Cartesianism

This chapter deals with collective tacit knowledge, which is the irreducible heartland of the concept. All that is needed here, however, is a brief introduction to the ideas, because comprehensive treatments can be found elsewhere.[1] What is new about this book is that I will relate what has been said about the social to what has been said about the body—somatic tacit knowledge—and what has now been identified as relational tacit knowledge. In previous treatments the significance of the social had not been properly separated from these other very different elements of the tacit.

Engaging in Social Life

Examples of the application of tacit knowledge discussed so far include bicycle riding, copy-typing, paint spraying, and chess. In each case it was argued that the knowledge needed to accomplish them, though not expressible in rules that could be executed by humans, was in practice, or in principle, expressible in rules that could be executed by machines. In the case of balancing on a bike and copy-typing, machines can already do the task better than humans, albeit by different means. In the case of paint spraying, at least some subset of the task can be managed. In the case of chess it depends on whether "playing chess" was taken to mean "winning at chess" or "winning at chess by using human-type pattern recognition." In the former case, as with bike balancing and copy-typing, machines can already do it better than humans; in the latter case, there are no machines that can do it but there seems no reason of principle why such machines could not one day be constructed. If there is an obstacle it would have to do

1. Collins and Kusch 1998 is an extensively worked-out examination of the problem.

with the affordance of the substance of human brain and body which may turn out to be difficult or impossible to reproduce by other means (like trying to make a slide rule out of elastic bands or a rocket out of ice).

In each of these cases, however, when they are carried out by humans in the normal course of social life, there are additional aspects of the activity to be taken into account that have not so far been discussed. The most straightforward case is bike riding. What was discussed in the last chapter was bike balancing but in normal human life bikes are ridden on roads used by others and traffic has to be negotiated; the gyroscopes and feedback systems that keep a mechanical bike upright cannot negotiate traffic. In the case of copy-typing, in real life there is always a trade-off between speed and accuracy. Where the task is done by machines this problem was finessed in the preceding chapter by positing that the source would be a clear and undamaged font—in other words, there would be no ambiguities. Real-life typists sometimes have to work with something less than this and they must make a judgment about how many mistakes to allow and how much time to spend correcting them and whether to reproduce or correct putative mistakes that are found in the original. Some of the abilities required are illustrated by the broken text found in box 15. In that case, the right number of corrections of mistakes in the original would, of course, be zero, but in the more normal course of events the typist has to make a judgment that turns on who and what purpose the work is being done for and what level of errors will be accepted in the short term and the long term—a social judgment. In spraying chairs with paint it is a matter of how thorough the job should be in different circumstances, how to adjust the thickness of paint according to the economy with which the job is to done, how to adjust time taken against quantity of paint used if the paint starts to run out, how to handle the job differently if the paint is loaded with a precious metal, and so forth. In world championship chess there is much more going on than what happens on the board—there is all the gamesmanship that begins with the setting up of the challenges and the location of the tournament—at which Bobby Fischer seems to have been an arch-exponent. Chess played under the circumstances of the cold war does not follow the same conventions of aesthetics and gamesmanship as chess in the first half of the twentieth century, and it has probably changed again with the victory of Deep Blue. There are no machines that can manage such trade-offs and repairs or apply the rules of gamesmanship in a human, social context-sensitive way. Nor is it possible to imagine how to invent such machines. To understand how these things are to be done we have to engage with social life.

Bicycling and Car Driving in Traffic

Now consider bicycle riding and car driving in more detail. Negotiating traffic is a problem that is *different in kind* to balancing a bike, because it includes understanding social conventions of traffic management and personal interaction. For example, it involves knowing how to make eye contact with drivers at busy junctions in just the way necessary to assure a safe passage and not to invite an unwanted response. And it involves understanding how differently these conventions will be executed in different locations. For example, bike riding in Amsterdam is a different matter than bike riding in London, or Rome, or New York, or Delhi, or Beijing. (For example, in China, bicycles are ridden at night, without lights, in ways that would be considered absolutely suicidal in the West.) Then again, even in one country there are different settings for riding—riding in the country and riding in the town—mountain biking, riding to school or work, racing, or riding as a display of skill.

Similar national differences apply to car driving. For example, in Italy a style of driving is adopted that passes responsibility for safety to other drivers—to the collective of drivers. Once this is first noticed one finds it is general for Italian drivers to expect the unexpected and cope with it as a matter of course while driving for the individual becomes much easier, since not everything you do has to be exactly "according to the book." This is driver collectivism. In Britain or America the individual must take much more responsibility for smooth traffic flow, violations from the rules of orderly flow being met with expressions of rage.[2]

Car driving in China is different again, collective responsibility being taken to an extreme. The Chinese drive on the right, but the following behavior, which I encountered on my first visit to China, was treated as unremarkable by the other (Chinese) passengers in the car: being driven along an elevated urban highway with two lanes of fast-moving traffic going in each direction, to save a few hundred meters when making his exit, the driver crossed the road to the left-hand side and went down the on-ramp, causing the vehicles coming up the ram to split and pass on either side. Likewise, passing can be accomplished on either side of the car being passed, irrespective of the width of the road; vehicles often use the hard shoulder and squeeze back into the smallest of gaps. Again, on any wide road in China all vehicles have a preference for the center and exercise it irrespective of speed or size of car. One bizarre consequence is that in the case of

2. See Collins 2004, ch. 22, for more discussion of these styles of driving.

right turns executed on wide and open roads with more than one lane in each direction, the central lane is held until the very last minute, then the turn is executed by cutting across the traffic on the inside lane, irrespective of speed. Finally, the notion of "cutting a corner" appears to have no application in China. Vehicles turning left routinely start by moving from the right-hand lane into the far left lane of the opposing carriageway in the face of oncoming vehicles. They make their turn in that left lane before moving back to the right side of the road after the turn is completed. Somehow, oncoming vehicles judge the intention and move out, leaving enough room for the turning vehicle to come through against the stream. The principle appears to be that right of way is given to whatever is there—pedestrian, bicycle, car, or truck having equal rights. It seems like madness but may be a stunning demonstration of how collective responsibility can be made to work—entirely appropriate in a Communist country. The key might also be the relative newness of dense road traffic in China. The mores would be entirely appropriate for slow, person-powered vehicles such as rickshaws, where saving a few yards is well worthwhile because of the physical effort needed to cover them.[3] These observations, by the way, were made in 2005—perhaps things have changed.[4]

What we have tried to do with these descriptions is to make sense of the very different driving styles found in different countries. But it should not be thought that the descriptions comprise a set of rules for driving in the countries described. In both Italy and China the overarching rule was "collective responsibility," but Chinese driving habits would not work in Italy, while in Britain and America if drivers resolutely ran over anyone who stepped into the road or crashed into any car that broke a rule, there would be chaos—so there exists a degree of collective responsibility even in these countries. What is required in every case is a social judgment about how individual responsibility and social responsibility are to be balanced and the right way to do things cannot be captured in any description on the page. The right way to do things can only be captured through experience, and that experience and its application vary from country to country. The explication of the way such things are captured through experience is the socialization problem.

3. Thanks to Neil Selwyn for making the point about rickshaws.

4. Though I was mostly terrified, over the course of a few days, the impression I obtained was, as intimated, one of a staggering degree of collective responsibility, something that might be thought worthy of envy. Unfortunately, the accident statistics show there is carnage on China's roads; and even in Italy, the accident rate is about double that of Britain and America.

Data

The difference between somatic and collective tacit knowledge may also be illustrated with the following fictional example (which can be treated as a thought experiment). In one episode of "Star Trek: The Next Generation," Dr. Crusher shows Lieutenant Commander Data—a clever android—the steps of a dance. She shows them only once. Most of us would need to practice and practice under Dr. Crusher's guidance before we could repeat the steps with her skill; for us it would be a case of acquiring somatic tacit knowledge. The scene is amusing, because this is what we are led to expect, yet Data is immediately able to repeat the steps at speed and without fault; he has the kind of quick brain that could also master bike balancing from Polanyi's instructions. If Data has a somatic limit, it is less restrictive than that of any human.

Dr. Crusher, impressed by the astonishing facility exhibited by Lieutenant Commander Data, then tells him that he must simply repeat what he has done with some additional improvisation if he wants to dance with verve on the ballroom floor. This is where "Star Trek" goes wrong, because it shows Data managing improvisation as flawlessly as he had managed the initial steps. But improvisation is a skill requiring the kind of tacit knowledge that can *only* be acquired through social embedding in society. Social sensibility is needed to know that one innovative dance step counts as an improvisation while another counts as foolish, dangerous, or ugly, and the difference may be a matter of changing fashions, your dancing partner, and location. There is no reason to believe that Data has this kind of social sensibility. Social sensibility does not come through having a quickly calculating brain, it comes through having the kind of brain that can absorb social rules. Data can follow Dr. Crusher's dance steps because his brain is so fast; tacit knowledge does not enter into his learning process. Data can master the steps after the fashion of a computer rather than in the clumsy way that humans would have to master them. But this won't work when it comes to the rules of improvisation.

Collective Tacit Knowledge and Two Phases of Skill Acquisition

Bike riding in traffic, car driving in traffic, and dancing all require learning with a degree of flexibility so that the style can be changed to fit different circumstances, such as riding or driving in different countries and dancing in different settings. Therefore, the change in the brain must be less fixed than in the cases of either weightlifters' muscles or those associated with learning to balance. The change must be malleable enough to respond to further changes in society and circumstances. This is a matter of condition 5. A corollary is that the change must happen in an entity that can be "in touch with" changing circumstances and somehow recognize the appropriate adjustments to what was initially transferred. It is this that presents the deep difficulty and unsolved mystery of the social aspects of tacit knowledge: by what mechanism do humans stay in touch with society and how can one build a machine that can do it?

Revisiting the Dreyfus brothers' five-stage model of skill acquisition, it is immediately striking that the step-change between bodily skills and social skills is not discussed. At some stage, "traffic scenarios" enter into the picture in the same way as engine sounds and gear-shifting once entered—just as more of the same. No cognizance is given to the fact that gear-shifting is going to be pretty similar in all countries—once learned in one country, the skill can be transferred to another (and we have the automatic gearbox)—while traffic scenarios are something different entirely (and we don't have the automatic "scenariobox"). Under the Dreyfus Model, there is no discontinuity in dancing such as has been exploited here by describing the "Star Trek" version of how Data learned. One can see how the discontinuity that has been described draws attention to the two phases rather than to five stages or even to the difference between conscious and unconscious execution. We have, instead of a five-stage model, two phases of skill acquisition (which are also the second two phases the Three Phase Model set out in this book). The first phase involves merely motor coordination while the second phase involves using that motor coordination in a socially sensitive way. Both phases may be conducted self-consciously or unselfconsciously, with the second usually being the more efficient. The Dreyfusian model has been very influential, but it reflects much less about the nature of knowledge itself than is usually imagined. It describes only one or two special kinds of human skill and it entirely ignores the most fundamental subdivision of human expertise: that between expertise of the sensory motor kind and expertise of the social kind. The one lasting division that is made, that between the self-conscious modality and the unself-conscious modality is orthogonal to this more fundamental division.

The Irreducibility of the Collectivity

What is being argued is that humans differ from animals, trees, and sieves in having a unique capacity to absorb social rules from the surrounding society—rules that change from place to place, circumstance to circumstance, and time to time. The difference between humans and other entities is most easily seen in the case of domestic animals. Domestic cats and dogs are exposed to human society as intensively as any human baby but just do not absorb the ways of going on. Obviously they do not learn human language, but they also do not learn the same distinctions between the clean and the dirty, the correctly placed and the misplaced. They never feel embarrassed about being naked nor about sniffing at each others' bottoms, or sniffing the private parts of humans, or having sex in public. They cannot be socialized. Furthermore, cats and dogs found in different societies do not have different mores. It is only humans who have the ability to acquire cultural fluency. It is only humans who possess what we can call "socialness"—the ability to absorb ways of going on from the surrounding society without being able to articulate the rules in detail.[5] As opposed to humans, there are no groups of vegetarian dogs, arty dogs, nerdy dogs, dogs that believe in witches, and dogs that understand mortgages—they are all just dogs. That one dog is different in "personality" from another dog is beyond dispute, it is just that these personality traits do not correspond to any significant cultural differences within breeds. Animals can all be understood as like bees, running on string transformations rather than languages; they are taught to transform (respond to) strings in the desired way by regimes of punishment and reward and/or evolution (like neural nets).

Social Cartesianism

What is being putting forward here is a strong claim about the existence of a radical difference between humans and other entities, including animals. The argument of the previous chapter was that there was no metaphysically significant discontinuity between humans and other entities if one considered only the individual human body and brain; in that case both humans and animals can be thought of as mere string transformers, though humans are the more complex and impenetrable. The difference between humans and other entities, tentatively introduced in the last chapter under the heading of "the self," turns on whether or not they can carry out polimorphic actions—actions that require different behaviors for successful instantiation depending on context and require different interpretations of the same behavior depending on context. The claim that this difference exists in a real sense will be called Social Cartesianism.[6]

Social Cartesianism claims that humans and animals are radically differ-

5. Collins 1998.

6. No other elements of Cartesianism are thereby endorsed.

ent. What it does not claim, unlike Cartesian Cartesianism, is that the boundary between humans and animals is sharply marked. There may be some animals, perhaps chimpanzees, perhaps cetaceans, perhaps birds, that share some human abilities in small ways. But this is not the issue. The issue is the marked difference in abilities between a species that possesses fully developed languages and cultures and one that does not. If there are borderline cases of animals that share small elements of language and culture, then these are curiosities but they do not affect the argument about the difference between having languages and cultures and not having them—they only affect the argument about exactly where the border is to be placed. It is necessary to belabor this point because many arguments concerning classifications take it that if it is possible to show that there is lack of clarity about the position of the border between two categories then those categories are not distinct. This is a fallacy. Thus, there is no exact boundary between a pond and field, with mud being an in-between category, but one may still drown in water but not in soil—the qualities of the two are quite different.[7] The same goes for the qualities of humans and animals irrespective of any fuzziness around the edge. All the many studies that purport to show that birds and chimps use language, or tools, or exhibit different behaviors in different groups, or learn different behaviors from each other affect the argument not at all.[8]

That understood, if Social Cartesianism is to be upheld, then a question that still has to be faced is whether there are equivalents of low-gravity bike riding in traffic, car driving in Beijing, dance improvisation, and so forth. In other words, would it be possible to write sets of instructions that could replace the tacit social understandings that underpin fluent performance of these activities if only we slowed time to a crawl? Fortunately, the question has already been debated, though the topic was seen as understanding rather than the artificial reproduction of social rules. The most well know

7. Searle (1969, 7) argues that even to be able to say what could *not* count as defining criteria—and it is easy to do this in the case of both pond and field: "a pond is full of earth," "a field is full of water"—we have to know what the things we are talking about mean. (He discusses the abstract notion of "analyticity," but his argument works just as well for ponds and fields.) Searle's argument seems related to the idea of social rules—we may not be able to state them but we know enough about them to say what they are not.

8. For an easily accessible summary of recent research along these lines, see Kenneally 2008. Another approach that does not bear on the argument here is that concerning "companion species" (Haraway 2003). I can have a companion dog, companion microbes, or companion blood cells, but my relationship with them remains asymmetric, just as my relationship with my hammer or my slide rule.

example of this debate is John Searle's Chinese Room. If the Chinese Room worked, then languages could be reduced to strings.[9]

The Chinese Room as a Low-Gravity Environment

The Chinese Room contains a large lookup table. On left side of the table are (supposedly) all questions that can be asked in Chinese; on the right side of the table are all the answers. Someone approaches the Chinese Room and passes a question written in Chinese through the letter box. A non-Chinese speaker inside the room locates the question on the left side of the table by matching the shapes of the characters and passes out the written answer found on the right side without knowing what either means. The questioner considers their question to have been answered, but Searle's point is that no one has had to understand Chinese to provide an answer (in our language, nothing but string transformation has taken place). Searle correctly deduced that the mere reproduction of an action, including one as complicated as answering certain questions in Chinese, does not demonstrate that consciousness or understanding or meaning is involved.

But Searle's problem is not ours. Our problem is whether the Chinese Room is a "low-gravity" way to speak Chinese. In other words, does the Chinese Room provide the conditions under which humans, who normally accomplish fluency in language only through internalizing tacit knowledge through socialization, could accomplish something indistinguishable from fluency by referring only to a lookup table and doing nothing more than string transformation. Or to put it another way, in contradiction to Social Cartesianism, can the normally polimorphic action of answering questions be mimicked with mimeomorphic actions? Though the Chinese Room does no more than string transformation, is it doing something that, in some miraculous way, is indistinguishable from language use, thus removing the mystery of collective tacit knowledge? If not, what has gone wrong?

An initial answer might be that there is an indefinite number of questions that can be thought up, so it would be impossible for the Chinese Room to contain all of the answers. But there might be a way round this problem implicit in an idea put forward by Ned Block.[10] To make things a little more tractable let's switch from a Chinese Room to an English Room—which makes no difference in principle—and let us start by assum-

9. See Searle 1980 and, for the basic critique to be developed here, see Collins 1990, esp. ch. 14.

10. Block 1981.

ing that questions have to be of a limited length. A question then would be something like this: "Please tell me what you like to do in hot weather and in cold weather and which of these you prefer?" This question is around 100 characters in length. Let this be the question length and allow the answer to be the same length, making a question-and-answer of 200 characters. Now we compile a list of all possible combinations of 200 characters constructed from what can be found on the typewriter keyboard (about 100 possibilities if we include both upper case and lower case, spaces, digits and all the punctuation marks). The resulting list contains 100^{200} items—that is, 10^{400} items. This compares with the 10^{125} particles in the known universe. That is to say, the number of particles in the universe is miniscule compared with number of possible combinations of only 200 characters given a choice of 100 at each location. We know then that the task cannot (*logistic principle*) be done, but let us not worry about this limitation.

From the list we pick just those strings that comprise reasonable questions and answers and assemble them into a lookup table. This list will be small compared with the 10^{400} items, though it will still be pretty large. One reason it will be a lot larger than it might be is that the question side of the lookup table, will have to include all the misspellings and other mistakes that a typical typist might well include in a typed version of a question (or a person such as myself might include to demonstrate the abilities of humans to repair broken text), and the answer side would have to include the mistakes that might be made by the responder if this machine is to reproduce the language capacities of a human. The possible number of combinations of "repairable" mistakes is vast. Therefore it is likely that it could still not be written down for reasons of *logistic principle* or, at least, *logistic practice*, but again, let that difficulty be ignored. What we have done, if you like, is build a machine that is the equivalent of a perfect brute-strength chess player. We have created a device that contains all possible question-and-answer games and therefore will always do the equivalent of never losing. Both the exhaustive brute-strength chess player and the exhaustive brute-strength conversationalist are beyond the realms of logistic possibility but we have decided to ignore that.

Have we now found a means of listing all possible questions and answers in a language and, if the resulting lookup table was deployed in the Chinese (or English) Room, have we found our low-gravity method of reproducing linguistic fluency? The answer is still "no."

What we have found, were it not logistically impossible, would be a way of converting a frozen moment of the language into a string. It is just a complicated, interactive version of the recipe for coq au vin discussed in chapter 1. It is a dictionary with bells and whistles. Even if it could be done as described, there would remain the problem that, first, sensible questions and answers are sensitive to context and, second, language changes.

The context problem is nicely illustrated in a lighthearted way by television comedienne Victoria Wood. Wood pretends to administer a questionnaire covering a series of disconnected topics to a passer-by in the street. She finishes with three questions:

> In the event of a nuclear holocaust, do you think there will be survivors or no survivors?
> Given that there were survivors, do you think there would be a total breakdown of society or some semblance of law and order?
> Given some type of structured postnuclear society, do you think people are more likely or less likely to be eating Hellmann's mayonnaise?[11]

Questions and answers in real life are structured by the unfolding context and, as Wood's sketch illustrates in an indirect way, that context is provided by both the conversation and what is happening in the world. The sketch reminds us that people would be unconcerned about the brand of mayonnaise they consumed immediately after a nuclear war so a Chinese/English room operating after such a cataclysm would have to change its repertoire if it was not to look like an idiot. So, unless the Chinese/English Room, or any more elaborate version of it, is sensitive to what is going on outside, it will be easy to distinguish from a human speaker of the language. This is the socialization problem.

The problem with the room is also easy to see if we think about the way language itself changes. If, in the 1940s, we had built a successful English Room it would not work today—the language and the vocabulary have both changed. Unless someone with good social sensitivity continually updates the database of a Chinese Room, it will soon become noticeably out of date, being ignorant of new usages and with a tendency to use archaic terms. But if the room is continually updated by someone with social sensitivity then it has not, in itself, solved the problem of social sensitivity. Once more, one can see that the best it can do, if it can do it, is to capture the language of a frozen moment.

The point is that any representation of the language in the transmission stage of communication is a set of strings. However elaborate the mechani-

11. "An Audience with Victoria Wood," DVD, directed by David G. Hillier (1988; Granada Media, 2002).

cal interaction of these strings, they remain strings unless they are embedded in the sea of interpretation that makes them a language rather than a set of transformations. All this is just to say, in a rather complicated way, what is expressed, much more briefly, by Socrates in *Phaedrus* (see the beginning of chapter 1). What Socrates says applies just as much to the Chinese Room and its variants as it does to any other book.

The same argument applies to every kind of action that turns on sensitivity to social context. Consider again the example of rule following first mentioned in the introduction—knowing how much distance it is proper to maintain between yourself and another passerby on the sidewalk. It depends on the country, whether it is night or day, whether they are of the same or opposite sex, whether the sidewalk is crowded or empty, whether it the city or the country, what part of the city it is, and so on. By collecting all past instances of one person passing another on the sidewalk, we could assemble a lookup table that contains a set of rules for passing someone on the sidewalk. But there still remains the problem of determining the current context and being sensitive to changing context. Walking on the sidewalk is, of course, just another version of car driving, bike riding in traffic, dancing, and the like. In all such cases, a lookup table is, again, frozen history at best. Changes in society are almost entirely beyond prediction and control outside of regimes such as that envisaged in George Orwell's *1984*.

To repeat, it could be argued that someone could be always at hand to update maintain the rule-base of something like the Chinese/English Room to take account of the changing society, but in that case all the social intelligence would be in the attendant because it would be the attendant who had the responsibility of recognizing the continuously changing contexts and conventions of fluency in the native language; as far as the pretense to social life was concerned, the machine would be, as it were, the attendant's marionette. Such an arrangement proves the point that it requires a human in the chain if the knowledge embedded in the society is to be transferred to the machine. It takes a human to socialize the machine because we have not yet solved the socialization problem. Neither the Chinese Room nor any elaboration of it is equivalent to a brute-strength chess player, because in chess the rules do not change whereas in language they change continually in response to the way society changes.[12]

12. See Mackenzie 2008 for an exceptional case in which even the rules of chess are not completely fixed.

The Metaphysics of the Collectivity

To work with the social aspect of tacit knowledge it is most parsimonious to adopt the perspective that the collectivity, rather than the individual, is the location of the knowledge.[13] This is another way in which Social Cartesianism differs from Cartesian Cartesianism. In Social Cartesianism, the individual is not the unit of analysis: the individual merely shares the collectivity's knowledge. The special thing about humans is their ability to feast on the cultural blood of the collectivity in the way that fleas feast on the blood of large animals. We are, in short, parasites, and the one thing about human brains that we can be sure is special is the way they afford parasitism in the matter of socially located knowledge. Neither animals nor things have the ability to live as parasites on social knowledge.[14]

Collective tacit knowledge is like somatic-affordance tacit knowledge in that it depends on the specific features of humans, per se. In the case of somatic-affordance tacit knowledge it is the material of the body and brain that is special—as the material of a slide rule is special (it has to conserve its shape). Note that in the case of somatic-affordance, the substance is pretty well continuous with the substance of animals; animals can learn to do most of the things that we can learn to do *as animals*. Animals can balance on bikes if well enough trained and where they fail, in the matter of, say, recognizing patterns on a chess board, in seems to be a matter of quantity, not quality—after all, animals are pretty good at recognizing other kinds of patterns. With collective tacit knowledge there is a discontinuity.

The Individual and the Collectivity

It may seem odd to say that the location of anything is the collectivity rather than the individual. It has been a fundamental notion in sociology at least since Durkheim but the bias toward individualism today is so strong in most academic communities that it is hard to posit a collective location for anything. Nevertheless, as intimated in chapter 5, there is nothing strange about this even for the most adamant materialist, because

13. The "founding father" of the idea of the collectivity, going as far as to talk of "collective consciousness," was Emile Durkheim (1933). For a (nowadays) rare invocation of the idea of the collectivity, see Knorr Cetina 1999, 178. For a discussion of the role of the essentially human agent in distributed cognition see Giere 2006, especially the second half of chapter 5.

14. Michel Serres has written on "the parasite" but it has nothing do with what is being discussed here.

there is simply nothing special about the boundaries of the skull or the skin.[15] My brain's connections do not stop at the boundary of my head, because they are not limited by the connections found within the grey matter; my brain's neurons are connected to the neurons of every other brain with which it is "in touch" via my five senses. There is nothing even remotely strange about saying that the seat of knowledge is the collectivity of brains because the collectivity of brains is just as much a "thing" as my individual brain is a "thing"; my brain is a collection of neurons separated by (if we were to examine them on an atomic scale) huge distances so the distance between brains in the collectivity is no obstacle to their comprising one "thing" between them. The collectivity of brains is just a large-scale version of my brain—it is just a bigger collection of interconnected neurons—and, as with synapses, the weights of the connections change whenever social and technological life is rearranged. So, if we don't like the metaphysics of the collectivity we can still accept the idea that knowledge is located in the collection of brains while remaining philosophically conservative. We can even say that the tacit knowledge that is associated with speaking language is located, not primarily in the individual brain but in the collection of brains. Interestingly, the very concept of the neural net shows us how to think about it this way without invoking anything mysterious like "collective consciousness." The metaphysically bashful can just think of all brains linked by speech as making up one big neural net.

The enduring mystery, as explained above, is just how I connect into the collectivity; a century of studies of childhood have not solved the socialization problem. For example, in spite of this enormous effort we still do not know how much of language is learned and how much is, as Noam Chomsky argues, innate. If it is still possible to argue about such fundamental things, the mechanisms must still be obscure.

Another reason it is so hard to think of the collectivity as the location of tacit knowledge is the sheer competence of the individual. Shut an individual in a room in the morning, and in the evening they will still be a fluent natural language speaker. But it is important to remember the problems with the Chinese/English room. The competence shown by the isolated

15. I take this formulation from Andy Clark and his theory of the extended mind. The point is, however, that the mind extends not just to tools and animals, as in Clark's model but, crucially, to other brains. Hutchins's notion of extended cognition (1995) is related, but the famous example of ship navigation seems more a matter of division of cognitive labor than collective cognition.

speaker will not last indefinitely; the individual is a temporary and leaky repository of collective knowledge. Kept apart from society for any length of time and the context sensitivity and currency of the individual's abilities will fade. H. G. Wells had it right in his story "The Country of the Blind," in which the language of a whole groups changes in response to their lack of contact with the wider society:

> For fourteen generations these people had been blind and cut off from all the seeing world; the names for all the things of sight had faded and changed; the story of the outer world was faded and changed to a child's story.[16]

Acquiring Social Knowledge through Words and Things

We have, then, a picture of the individual as a parasite on the social group, sucking up social knowledge from the super-organism; stop sucking and the knowledge gradually degrades—that is, its match with the collective's knowledge gradually weakens. But how does social knowledge pass into the individual? So far we have simply said that it happens as a result of "being immersed" in society.

Immersion is participating in the talk and practices of society. A closer examination of the way we are immersed in society sheds a little more light on the relative role of body and collective. The Sapir-Whorf hypothesis holds that the words in a language depend on the physical surroundings; the old cliché is that Eskimos have seventeen words for snow, because snow comprises so much of their business. Whether this is really true or not does not matter, because it is certainly true in essence. If your people have grown up isolated in the Amazon jungle, your native language almost certainly has no word for snow, or house brick or bicycle or car. Globalization aside, a natural language will be in part a function of the things that are around. One type of thing that is around, are the bodies of the speakers. As it happens, if we forget about the "Country of the Blind," every people that speaks a natural language possesses a similar body type.

Body type affects language in three ways. First, physiology creates the very conditions for language. Language begins with speech. Without a cer-

16. Wells 1904, 854. One might illustrate the point further with the metaphor of the immune system: however well prepared a child is for the biological environment via the antibodies in his mother's milk, isolate him from dirt and his immune system will start to fail—he will no longer be ready for interaction with the changing world of infective agents.

Box 17. Wittgenstein and the lion.

"One human being can be a complete enigma to another. We learn this when we come into a strange country with entirely strange traditions; and what is more, even given a mastery of the country's language. We do not *understand* the people. (And not because of not knowing what they are saying to themselves.) We cannot find our feet with them. . . . If a lion could speak we would not understand him." (Ludwig Wittgenstein, *Philosophical Investigations* [1953], quoted on p. 223of the 1967 Oxford edition)

tain kind of larynx, a certain kind of brain, and certain kind of lung, all of which probably coevolved, there would be no natural languages—at least not natural languages as we know them.[17]

Second, the body determines the standard kind of string that is used in communication without artificial aids. For example, inscribed alphabets and icons will have elements of a certain mean size and conversations will be held around a certain mean volume adjusted according to how many people are meant to hear—these amplitudes are a function of physiology. This is not a profound point: all animal species are roughly equal in their hearing and discerning abilities so we can treat this feature of physiology as background (like the fact that all animals have blood).

Third, the shape of the body affects the terms in the language and, therefore, the conceptual structure of the world. A standard example, already mentioned in the introduction, is "chair": the fact that humans walk on two legs and have knees that bend backward enables them to take comfort from chairs. If we walked on four legs we almost certainly would not have the concept of, or word for, "chair." The relationship between my body and my conceptual world is captured in Wittgenstein's famous remark about our not being able to understand a lion even if it could talk (box 17). My conceptual world has been made for me by embedding in a species which does not have the right teeth, claws, or nose to equip it with lions' concepts; likewise, lions, presumably, cannot make sense of the idea of a chair.

17. Of course, it is not hard to imagine that signing might have developed as an initial natural language in a people without vocal chords and ears. Sign languages have a different conceptual structure to natural languages.

The Social and Minimal Embodiment Theses

The above arguments seem to give the body great importance even in the life of the collective. They seem to suggest that there can be no stark contrast between somatic tacit knowledge and collective tacit knowledge, because knowledge located in the collectivity is, to a considerable extent, formed by the body. Furthermore, since being a conceptual parasite on society usually involves participating in the practices that form a society, it seems that the body is necessary to the acquisition of even collective tacit knowledge.

What is certainly true is that the content of collective tacit knowledge is a product of the typical shape of the body of the species that comprise the collective. The content of the collective tacit knowledge of speaking lions, if they existed, would look different than the content of the collective tacit knowledge of humans. And, that humans have tacit knowledge and non-talking lions don't is again a matter of the way our bodies are—our species' brains and larynxes. This is the *social embodiment thesis*.

But to understand ourselves as social parasites we have to be clear that not every individual needs the typical body in order to draw on collective tacit knowledge. This is because collective tacit knowledge is, to a large extent, located in the language of the collectivity rather than its practices. And remember, becoming fluent in a language is not like becoming a Chinese Room or equivalent; it is to master the tacit knowledge inhering in the conceptual life of a society.[18]

While the collectivity's language can only develop its characteristics as a result of its members' practices, one can be a parasite in respect of that language without engaging in the practices. And one can be a parasite in respect of the language with no more than the minimal body required to engage in discourse—the kind of brain that is associated with the language-speaking species and the larynx, lungs, and ears or prostheses that can take their place. Thus, Steven Hawking can still acquire new elements of collective tacit knowledge while remaining perfectly still and using prosthetic devices to speak. "Madeleine," described by Oliver Sacks as born blind, and disabled, and unable even to use her hands to read braille, had "never fed herself, used the toilet by herself, or reached out to help herself, always

18. Wittgenstein, in his famous remark about the lion, allowed himself to make a mistake in this respect, as can be seen in box 17. There he talks of someone first entering a country with mastery of the language yet not understanding the people. Mastering the language *means* understanding the people.

leaving it to others to help her," could still become a person who "spoke freely indeed eloquently . . . revealing herself to be a high-spirited woman of exceptional intelligence and literacy" through the medium of the written and spoken word.[19] This is the *minimal embodiment thesis.*

The difference between the social and the minimal embodiment theses provides a stark contrast between humans and animals: the difference applies in humans but not in animals. An animal's body is an individual thing. A rabbit born without legs and eyes will never know what it is to have a rabbit-like body. But a human born without legs and eyes can know what it is to possess the collective human body shape; a human can share the knowledge through the medium of a language that has been part formed through the physical interactions with the world of all those other human bodies and carries with it the conceptual structure of the fully embodied world.

To acquire collective tacit knowledge without engaging in or having the ability to engage in collective practices is known as acquiring interactional expertise.[20] It has been hard to establish a case for the existence of interactional expertise, partly because of the recent overvaluation of the role of the body in conceptual life among philosophers and others, probably reacting to the massive overvaluation of the power of logic and reason in the 1950s. It is also hard to establish the case, because except for unusual people such as Madeleine, linguistic competence is nearly always acquired alongside practical competence, so the two are hard to disentangle. Recent research has shown that they can be separated, however. There are instances of socialization where words predominate, and in these cases interactional expertise does not seem to be limited by lack of practice or practical competence. The case of the congenitally disabled has already been mentioned, but the same absence of practical immersion is found in more humdrum circumstances. Anthropologists, ethnographers, or interpretative sociologists immerse themselves for shorter or longer times in the society they are studying, sometimes without engaging much in the practice but still acquiring cultural fluency. I am an instance, having spent decades immersed in the gravitational wave detection, written an 875-page book on the field, and managed to pass an "imitation game" test as a gravitational wave physicist; I did these things without making any significant contribution to grav-

19. Sacks 1985, 56–59.

20. See Collins and Evans 2007 for a full discussion of interactional expertise and many other categories of expertise. See Collins 2007b for more applications of the notion of interactional expertise.

itational wave detection experiments or the writing of gravitational wave papers—the central activities of the gravitational wave detection community.[21]

In the same way, managers of large scientific projects have to make "informed" decisions about the value of different technological pathways without themselves ever having taken part in research practice of the relevant specialties. For example, Gary Sanders moved from being a researcher in high-energy physics, to becoming the project manager of the Laser Interferometer Gravitational-Wave Observatory and then the project director of the Thirty Meter Telescope (TMT); he reported that a crucial aspect of his abilities was interactional expertise.[22] Indeed, any scientist working in a large collaboration has to engage in the discourse of other specialties without having engaged in the narrowly defined practices if the collaboration is to be more than an aggregation of separate individuals—technical discourse has to be learned outside of the technical practice in this kind of situation. In fact, one can argue that interactional expertise is necessary to have any kind of division of labor in society which goes beyond slaves answering to the whip.

Of course, drawing on the tacit knowledge of the collectivity through language alone is often not the most efficient way to do it. Engaging in physical activity with other people tends to create more opportunities for conversation than engaging in talk alone, so even if the sole means of transmission was the word, the transmission would be enhanced by physical activity. Furthermore, practical competence, or the attempt to master practical competence, is usually what gains entry into the locations where linguistic fluency is learned. For these reasons, a person who has taken part in both conversations and practical activities is likely to be further ahead in the acquisition of collective tacit knowledge than a person who has been exposed to words alone.[23] On the other hand, physical skills, such as those involved in making contributions to gravitational wave physics, may be too hard or take an inordinate amount of time for someone like a visiting sociologist to master. Likewise, it is too much to ask that all the skills pertaining to the many scientific specialties that the manager of a big scientific project or the specialists within it need to understand if the project is to run in a coordinated way, all be acquired via practice. The fact that interactional expertise,

21. Collins 2004 is the long book; discussion of imitation games and of the particular test can be found in Collins and Evans 2007.

22. Collins and Sanders, 2007.

23. As analyzed by Ribeiro 2007.

once one starts to notice it, is "everywhere" attests to the fact that it is, in many circumstances, the most efficient means of acquiring collective tacit knowledge.

Could All Collective Tacit Knowledge Be Made Explicit?

The fact that acquiring collective tacit knowledge via practice is sometimes more and sometimes less efficient than acquiring it through discourse alone is, once more, a matter of the nature of humans not the nature of knowledge. No human body is good at acquiring lots of practical skills and some human bodies are better at acquiring some practical skills than other human bodies. But these different efficiencies are not "metaphysically significant," as one might say. As far as knowledge is concerned, the deep mystery remains how to make explicable the way that individuals acquire collective tacit knowledge at all. We can describe the circumstances under which it is acquired, but we cannot describe or explain the mechanism nor build machines that can mimic it. Nor can we foresee how to build such machines in the way we can foresee how we might build machines to mimic somatic tacit knowledge. In the second case we know what we would need to do to make them work, in the first case we will not know how to start until we have solved the socialization problem.[24]

Again we ask the standard question: could all collective tacit knowledge be made explicit? The answer is "no"—at least, not in the foreseeable future. It cannot (*technological impossibility or higher*) be done.

The argument of this chapter is that the central and still mysterious domain in the map of tacit knowledge is knowledge that is located in society. It has been argued that we cannot foresee how to make machines that have the ability to be social parasites, nor to describe the ability scientifically, nor to discover long strings that will encapsulate the ability. It is crucial to see that there is a difference in the role of the typical body and that of the individual body—it is important, in other words, to understand the minimal embodiment thesis. To be a social parasite only a minimal body is necessary but the human brain is vital. The complete bodies of nonhuman animals are useless when it comes to being a social parasite.

24. Attempts to build socializable machines, such as Cog or Kismet, are a kind of "cargo cult science." Something with some slight resemblance to a human is constructed in the hope that, in some miraculous way, it will acquire human abilities without needing to understand those abilities first.

PART III

Looking Backward and Looking Forward

SEVEN

A Brief Look Back

In this chapter some existing approaches to tacit knowledge and some previous case studies are briefly reexamined in the light of the preceding arguments. To reiterate remarks made in the introduction, I make no attempt at a literature review but merely pick a few items that are indicative of the relationship of this work to the broad themes of some previous approaches; it is a matter of clarification.[1] A critical look at some of my own studies completes the exercise.

Economics and Management

In the economics and management literature, the division between tacit and explicit knowledge is taken to be unproblematic. To take one example, Shantanu Dutta and Alan Weiss write that "codified knowledge is amenable to the printed page and can easily be transmitted, such as in designs and specifications, and is therefore less proprietary than tacit knowledge (Polanyi 1958), which is far more difficult to codify and hence difficult to imitate."[2] If we are to retain the contrasting notions of explicit and tacit knowledge, then this sentiment reflects experience in the world as it is encountered, but that is only because the world as it is encountered is full of people with the shared background of tacit knowledge that allows "designs and specifications" to afford meaning—that is to say, to work as conditions 1, 2, and 3 communications without needing the changes to the recipient

1. See Gourlay 2006b, Hedesstrom and Whitley 2000, and Whitley 2000 for discussions of a wider range of empirical and other approaches to the notion of tacit knowledge.

2. Dutta and Weiss 1997, 345. Thanks to Rodrigo Ribeiro for bringing my attention to this literature.

that are associated with conditions 4 and 5. The apparent sharp difference described by Dutta and Weiss is, then, a matter of the world as we know it but it does not map on to an ontological divide between the tacit and the explicit.[3] To treat this difference as an ontological divide is to treat something hard to understand as something easy and to risk going wrong when "designs and specifications" have to cross cultural boundaries.[4]

Many of those who encounter the concept of tacit knowledge through the management literature will find that Ikujiro Nonaka and Hirotaka Takeuchi's study of the bread-making machine is treated as a "paradigmatic" application of tacit knowledge. But Nonaka and Takeuchi's description of the concept of tacit knowledge is likely to be useless for anything except the narrow purpose they have in mind.

Nonaka and Takeuchi describe the process by which a mechanical bread-making device was put onto the market. The key mechanism, as they describe it, is the transformation of tacit knowledge into the explicit knowledge needed to set out the design for the machine. "Under this theory, tacit knowledge is reduced to 'hidden' or 'not-yet-articulated knowledge' waiting to be uncovered and explicated."[5] For example, the ability to knead dough successfully required that the description "twist and stretch" be elicited from the baker, but once expressed in this way there was no further deep obstacle to mechanizing it. Nonaka and Takeuchi's description of the bread maker has been widely criticized, but the particular criticisms mounted here arise, first, out of the framework of this book and, second, out of observations of the actual process of making bread.[6]

According to the analysis presented here, the tacit knowledge involved in breadmaking by a master baker is likely to consist of some relational tacit knowledge, some somatic tacit knowledge, and some collective tacit knowledge—RTK + STK + CTK. The articulation of the "twist-stretch" element of breadmaking is an example of a piece of relational tacit knowledge becoming explicit. Whether it, on its own, would afford the execution of kneading we do not know: it might do or it might comprise an instruction equivalent to Polanyi's formula for riding a bike beyond our somatic limits. In that case the actual execution of the twist-stretch by a human baker

3. This is a mistake made even by a philosopher as sophisticated as Dreyfus—see the discussion of the supposed "knowledge barrier" in Collins and Kusch 1998, 194.

4. Ribeiro 2007.

5. See Tsoukas 2005, esp. p. 154.

6. As described in Ribeiro and Collins 2007. Existing critical analysis of Nonaka and Takeuchi include Tsoukas 2005 and Gourlay 2006a.

would involve somatic tacit knowledge. In the case of the mechanical bread maker, however, it was not difficult to find other materials with the right affordances to do the job and certain advantages in terms of endurance (as with bicycle balancing machines).

Just as in the case of the bicycle, however, there is a raft of collective tacit knowledge that is required to bake bread in the real world, and this varies from country to country and setting to setting. In the real world bread is made with huge variations of ingredients, in a huge variety of sizes, shapes, colors, and crusts. And bread is made with more or less care according to the anticipated consumer and the expected pattern of consumption. In line with the thesis of this book, the automated breadmaker is a complete failure in respect of these aspects of bread making—the aspects that involve polimorphic actions.

Some approximation of human bread making can be achieved even in respect of collective tacit knowledge if the polimorphic actions are substituted by mimeomorphic actions. Thus the machine is designed to offer a few selected choices of degree of baking and finish, which can be selected by pressing the appropriate buttons and some different styles of bread can be chosen by purchasing different packs of pre-prepared ingredients or by following a choice of recipes. The choices still have to be made but the range of choices is limited so that the execution of the choice is mimeomorphic.[7]

Even this does not exhaust the story of the various kinds of tacit knowledge, however, for both breadmaking by hand and automated breadmaking are set in their action trees of other kinds of activity. The human bread maker has to know how to purchase and measure ingredients and use the bowls and the ovens, while the user of the mechanical bread maker has to know how to read and follow instructions, purchase ingredients, clean the machine and use it and repair it, or get it repaired, if necessary. To understand this it is only necessary to try the standard thought experiment of imaging oneself taking the automated bread maker to a tribe in the Amazon jungle and presenting it to them as a solution to their breadmaking problems. All these wider aspects of breadmaking will themselves turn on a mixture of the three categories of tacit knowledge and one can see how, in the case of the Amazon dwellers, only extensive condition 4 and 5 transfers—not to mention a transformation in the technological infrastructure of the

7. To adhere to the language of Collins and Kusch 1998, it is a "disjunctive mimeomorphic action."

jungle world—would enable the mechanical bread maker to be used successfully.[8] In these respects, the mechanical bread maker is a social prosthesis, just as the pocket calculator is a social prosthesis. It works only because the surrounding social organism makes up and "repairs" its deficiencies. These repairs usually take place without anyone noticing, which is probably why the standard analysis of these things goes so wrong so often.

In sum, Nonaka and Takeuchi, like so many others, mistake our surface encounters with the tacit and explicit as reflecting something deeper and, as a result, any attempt to use their approach for anything more ambitious is likely to end in failure. The kind of failure to be expected is like that of the Japanese attempt, begun in 1982, to build a "fifth generation" of expert machines.[9] I predict failure will also be the outcome of what has been said to be the current Japanese project to model the abilities of a five-year-old child by the year 2020. If the project was to model the abilities of a full-grown cat or dog, on the other hand, then, according to the arguments of this book, it might succeed. The child, by the age of five, will be a social creature drawing on collective tacit knowledge, but with even full-grown cats and dogs nothing more than somatic tacit knowledge is at stake. Of course, someone may make an unforeseeable breakthrough in the interim, but the argument of this book is only that it is currently unforeseeable. The prediction is offered as a hostage to fortune—a falsifiable claim.

Another interesting discussion of tacit knowledge that can be found in the management literature is Philippe Baumard's *Tacit Knowledge in Organizations*.[10] Baumard suggests that when things stop running smoothly in a company, codified knowledge, such as the organizational chart, can only stand in the way of a solution. In that case the company has to fall back on some kind of collective tacit knowledge. Unfortunately, it is not clear what Baumard means by collective knowledge and the like. But it is possible to read it as supporting the thesis of chapter 6, namely, that an organization has a degree of autonomy and somehow possesses knowledge upon which its members draw when they make decisions; they draw on the organizational knowledge without being aware that they are doing so and certainly without being able to express fully what they know as a result of their membership of the organization. Under this reading, an organization is like a form of life or a Kuhnian paradigm. We know that scientific para-

8. See Bijker, Hughes, and Pinch 1987 for a discussion of infrastructure and technological possibility.

9. http://en.wikipedia.org/wiki/Fifth_generation_computer.

10. Baumard 1999. I am grateful to Stephen Gourlay for discussion of this book.

digms provide scientists with both their basic assumptions and their experimental practices, the two being intimately related as concepts and practices are in Wittgensteinian forms of life.[11]

Philosophical Approaches

The central difference between the approach of this book and that of the most influential current philosophical approaches to knowledge shows up most clearly in chapter 5. For many philosophers the experience of human acquisition of knowledge is taken to be the principal exploratory tool in the analysis of knowledge as a whole. The body dominates human experience, and the result is that somatic tacit knowledge has unwarranted salience and is confounded with other kinds of tacit knowledge. Philosophers such as Heidegger (as interpreted by Dreyfus—see note 12 in chapter 5) and Merleau-Ponty typify the problem. In the same way, those who treat the study of human development—developmental psychology and the like—as the key to understanding knowledge can all too easily confound the categories because, in the normal way, all three phases develop together in the course of growing up. Developmental psychology offers valuable insights into the processes humans go through as they become parasites on collective tacit knowledge but the explanation is far from complete.

Another reason it is misleading to study human learning as a means of studying knowledge as a whole is that, as has been mentioned previously, humans encounter all three phases of tacit knowledge in the same way. It is often the case that even explicit knowledge is transmitted by close proximity between teacher and learner along with guided instruction. The same is certainly true of somatic tacit knowledge; the experience of learning how to balance on a bike or play chess well (or drive a car) is, in terms of personal experience, indistinguishable from learning a language or any of the other features of becoming a social being. In every case the transference is done by demonstrating, showing, and guiding, involving direct contact between

11. I am talking here of the notion of paradigm in its original formulation before Kuhn mistakenly watered it down in response to criticism. The strength of the idea of paradigm is the combination of concepts and practice, and if Kuhn had understood his own idea in the light of Wittgenstein he would not have separated the notion into the two parts (see Pinch 1982). This kind of approach to organizations is not entirely dissimilar to that found in the field known as "organizational learning." Schatzki (2003, 2005) has a more theorized notion of the organization based on what he calls "site ontology." To quote Schatzki, "the ultimate location of social life is sites composed of material, cultural and institutional practices" (personal communication, June 29, 2008).

someone who has the knowledge and someone who does not. It follows that for the student or philosopher of human *experience* there is no obvious distinction between the different regions of the map.

In setting out a critique of previous philosophical approach I concentrate closely on Dreyfus. This is because Dreyfus's aims are closer to the aims of this book than that of other philosophers of the body and because his ideas are always admirably clear; criticizing Dreyfus is likely to result in the greatest degree of clarification of the differences between the approach of this book and that of existing influential philosophies because Dreyfus is the most straightforward of the philosophers who deal with these matters.

Dreyfus can be criticized on three fronts. First, as a philosopher of the body he reads too much into the way humans acquire skills. For example, the careful analysis of car driving undertaken for the purpose of building the five-stage model obscures the much more crucial divide between somatic tacit knowledge and collective tacit knowledge (not to mention the fact that it applies to a limited range of examples).

The second criticism has to do with how this crucial divide is understood. Dreyfus knows that machines do seem to be able to do certain human tasks as well or better than humans. For example, he has noticed that computers are pretty good at mathematics and certain other science-related tasks. This he takes to be a consequence of what amounts to a "knowledge barrier."[12] Thus, computers, according to Dreyfus, can work well in areas that involve "the conceptual rather than the perceptual world" where "problems are completely formalized and completely calculable."[13] But this is careless. The close studies of science that have been conducted since the early 1970s show that there are no well-structured domains, nor areas which are completely formalized and completely calculable. Complete calculability is not an explanation; rather, it is the appearance of complete calculability that has to be explained. It can be explained by the existence of mimeomorphic actions on the one hand—areas of life where humans decide to act in such a way that their actions can be mimicked by machines—and our remarkable abilities to repair mistakes without noticing—to use machines as social prostheses while treating them as full members of society. Once more, the example of the pocket calculator is a good one to keep in mind. The key, as always, is to remember that most human activity involves the three tacit components—

12. For the term "knowledge barrier" as applied to Dreyfus's position, see Collins and Kusch 1998, 194. The same confusion is found in Herbert Simon's idea of "well-structured and ill-structured domains" (Simon 1969, 1973). There are no well-structured domains.

13. Dreyfus 1992, 291–93.

RTK, STK, and CTK—and whenever a machine appears to take the place of the human one must look closely to see how those three elements are being mimicked, substituted, or repaired. Once more, it is useful to try the thought experiment of taking the machine into an alien context—such as a tribe in the Amazon jungle—that makes the element of repair (the background of tacit knowledge, as Polanyi would say) stand out because of its absence.

The third criticism of Dreyfus is that he is so concerned with practice that he is unable to come to terms with the idea of interactional expertise and the difference between the social and minimal embodiment theses—that bodily form may determine the conceptual structure of the language of a species but it does not determine the conceptual structure of the language of an individual; the conceptual framework of individuals is given to them by the species or culture in which they are embedded—with the bridge being interactional expertise. Only with this idea in mind can we understand how it is that someone who has not engaged in an activity can still have a fluent understanding of it.[14]

Dreyfus continues to claim:

> You may have mastered the way surgeons talk to each other but you don't understand surgery unless you can tell thousands of different cuts from each other and judge which is appropriate. In the domain of surgery no matter how well we can pass the word along we are just dumb. So is the sportscaster who can't tell a strike from a ball until the umpire has announced it.[15]

But how can this be? Am I "just dumb" in the matter of murder unless I have committed it? Am I "just dumb" in the matter of politics unless I have been a politician? Am I "just dumb" in the matter of traffic accidents until I have been in one? I play bad golf and have never won a golf competition: do I have no idea what winning would feel like? If so, why might I want to win? It might feel really unpleasant! And so on. The notion that practical accomplishment is a necessary condition of conceptual grasp leads to absurdity.

It may be that Dreyfus's position is born of confounding what usually leads to the acquisition of linguistic fluency—access to the locations where linguistic fluency can be learned—with the methods by which linguistic fluency is acquired. The good football analyst is usually an ex-football player because only football players have been in the right position to engage in

14. Collins et al. 2006; Collins and Evans 2007.
15. Selinger, Dreyfus, and Collins 2007, 737.

lots of conversation about football with other people who can talk fluently about the practice of football. But access can sometimes be gained without practical accomplishment and then, as has been shown, the practical accomplishment is not a condition of the conceptual fluency.

The overvaluation of the role of the body and human bodily experiences as the key to understanding knowledge is almost certainly a reaction to the overvaluation of the power of logic and reason among, say, the artificial intelligence community. They tend to see the attainment of linguistic fluency in a machine as just a matter of more and longer strings. The repairs necessary in ordinary conversation, made so evident by the failures of spell-checkers and speech transcribers, are taken to be symptoms of a task not yet completed rather than of a phenomenon than cannot be currently understood—the way humans interact with society. The bulwark of bodily competence was erected to defend against the over-logicization of the world but times have moved on and we can afford to relax a little.

Polanyi

Michael Polanyi is the inventor of the term "tacit knowledge," and his remarks have been the jumping off point for some of the arguments in this book. But Polanyi was writing in the postwar period when the world was a very different place. In particular, it was a world dominated by science and the belief that logic, observation, and experiment were, without question, the preeminent ways of obtaining sure knowledge. As I argue in the introduction, in such a context, with such an uphill battle to fight, it is no surprise that Polanyi was tempted to make tacit knowledge into something mystical and inspirational (box 18). I believe that his stress on the personal element of tacit knowledge can do damage to the proper understanding of the idea, the profound parts of which have much more to do with the collective embedding of knowledge.[16]

Polanyi points out quite correctly that, say, even an act of measurement depends on a person's judgment about whether the measurement has been made to an appropriate standard and so forth.[17] Good judgments are a matter of experience and experience goes along with having tacit knowledge. The personal aspect of Polanyi's "personal knowledge," then, is not knowledge at all, but is the process of making good judgments and that arises out of hav-

16. The later philosophy of Wittgenstein is the better resource.

17. Polanyi 1975, 30. See also Kuhn 1961 on measurement and Collins 1985 on the notion of the "experimenter's regress."

Box 18. Personal knowledge.

"We must conclude that the paradigmatic case of scientific knowledge, in which all faculties that are necessary for finding and holding scientific knowledge are fully developed, is the knowledge of approaching discovery.

To hold such knowledge is an act deeply committed to the conviction that there is something there to be discovered. It is personal, in the sense of involving the personality of him who holds it, and also in the sense of being, as a rule, solitary; but there is no trace in it of self-indulgence. The discoverer is filled with a compelling sense of responsibility for the pursuit of a hidden truth, which demands his services for revealing it. His act of knowing exercises a personal judgment in relating evidence to an external reality, an aspect of which he is seeking to apprehend." (Michael Polanyi, *The Tacit Dimension* [1966], 24–25)

ing stores of tacit knowledge. Insofar as these stores have an element of collective tacit knowledge they link the person back to the society in which the judgment is embedded.[18] The ability to make good judgments is often referred to as "intuition," and that is a useful term so long as we remember that it is "wisdom based on experience"; this is mysterious enough.[19] This kind of "intuition" can be gained through practice and socialization, including the acquisition of interactional expertise. An attempt to analyze judgment of this kind and its relationship to tacit knowledge is made in *Rethinking Expertise*, which could, perhaps, be thought of as an analysis of the grounds of the personal judging part what Polanyi called personal knowledge.[20]

Critical Discussion of the TEA Laser and "Q of Sapphire" Studies

My study of the way scientists learned to build a new kind of laser, the transversely excited atmospheric pressure carbon dioxide, or, TEA, laser,[21]

18. See also Tsoukas (2005, 126–27) for a discussion of Polanyi's treatment of personal judgement and its relation to context.

19. Benner (1984) uses the Dreyfuses' five-stage model, including the intuitive aspect, in her well known work on nursing. For another nice study of the role of this kind of intuition in nursing see Herbig and Bussing (2001). See also Gladwell (2005) for the power of unreflective "intuition" in a number of areas.

20. Collins and Evans 2007.

21. Collins 1974, 1985. In this section I look back at two of my own empirical studies to see how they fit with the schema developed in this book.

showed that the scientists failed if they used only the information published in scientific papers. These papers included those which supplied details as intricate as the cross-section and machining instructions for the electrodes and even the manufacturers' part numbers for bought-in items. It showed, however, that only those who spent some time socially interacting with others who had already built a working model could succeed. The finding was described as supporting the "enculturational model" of knowledge transfer as opposed to the "algorithmical model": learning to build the laser was like learning a new language or culture, rather than absorbing discrete new pieces of information. The need for an enculturational model of learning could be said to be a consequence of the "rules regress"—the fact that rules can never contain all the rules for their own application.[22] My 2001 study of scientist trying to measure the quality factor, or Q, of sapphire, backed up the earlier work by showing that measurements of the quality factor of small sapphire crystals were so hard that only one group of scientists in the world were able to achieve them until a member of the successful Russian group spent considerable time in the laboratory of a second group, in Glasgow, who finally managed it after a week or so of interaction.

Both studies deal, *inter alia*, with relational tacit knowledge, though the category was not conceptualized (this book is the first appearance of the term). The TEA laser paper introduced the idea that scientists might keep secrets that could be revealed by a laboratory visit or that a laboratory visit might transfer information the importance of which is not known to either one or both parties. The sapphire paper breaks tacit knowledge into five categories: (1) concealed knowledge, (2) mismatched saliences, (3) ostensive knowledge, (4) unrecognized knowledge, and (5) uncognized/uncognizable knowledge. As can now be seen, the first four categories are different types of relational tacit knowledge (ostensive knowledge amounts to using artifacts as extra-long strings—see chapters 2 and 4). Uncognized/uncognizable knowledge, which is not explored in any detail in the sapphire paper and is alluded to in the laser paper, could be either somatic tacit knowledge or collective tacit knowledge.

Both papers offer a number of examples of the different kinds of knowledge that could only be passed on through personal contact rather than intermediary persons or things. They show, then, why intermediary persons or things could not do the job and why the knowledge that was passed on

22. Illustrated in Collins 1985 with the game Awkward Student, which turns on continuation of the series "2, 4, 6, 8."

during a laboratory visit was properly described as tacit knowledge. The sapphire paper discussed the conditions under which receivers of information would think it worthwhile to do the hard work necessary to learn tacit knowledge—trust in the integrity of the sender being the crucial factor. Neither paper broke down the nature of the knowledge being transmitted into interactional and contributory but that distinction was not made until 2002. Though both papers can be partially redescribed in the language of this book, the redescription does not change the main conclusions that were drawn at the time.

The papers, like much of the existing literature of tacit knowledge, work in the narrow context in which they are set—they show that to learn from others how to build a TEA laser or measure the Q of sapphire, written descriptions are not enough and personal contact is needed. The papers show, with detailed examples, why this is the case and one way to classify the reasons. It is when we turn to bigger questions that more can be understood by considering the matter in the terms of the argument presented here.

Consider the TEA laser case. What would now be said is that the strings available in the printed literature on TEA laser building were too short to afford the interpretations necessary to make a laser work when passed between those who knew and those who did not know how to build a laser. The paper made it clear that even strings as long as that represented by articles published in the very detailed *Review of Scientific Instruments* were insufficient to the task. What we can now say is that though longer and longer strings do not seem to have resolved the problem in this case, they might have done should they have been long enough to from a bridge between the tacit knowledge of sender and receiver. Instances of this were described. For example, some laser builders' failures were resolved when it was found that the top lead must be no longer than eight inches. In 1972, nobody knew this but those who followed existing designs automatically built with a short top lead, whereas those who followed a circuit diagram were likely to use a longer lead. When the importance of this serendipitous feature of the design became understood it could be explained to novice builders so some tacit knowledge had been explicated. Thus, a condition 1 communication had failed (long though the strings were), while a condition 3 communication, with slightly longer strings, succeeded in respect of one aspect of the design.

It should be expected that the knowledge required to build TEA lasers would become more widespread over time; the culture of laser building would become more ubiquitous, as we might say. In that case, shorter strings would be likely to work, because a lot of the knowledge would be absorbed during the normal course of socialization of a physicist. The condition 4 or condition 5 changes mean that condition 1, 2, and 3 communications would become more likely to succeed. The normal course of socialization involves the acquisition of collective tacit knowledge. Doubtless, the actual putting together of a laser also requires bodily skills so, once more, building a TEA laser is a matter of RTK + STK + CTK.

An interesting question thrown up by the current treatment is how to describe what would be going on if, given the situation pertaining in 1974, a laboratory with a working laser were to pack it in a crate and deliver it to a laboratory that, up to that point, had not been able to make a laser that worked. In terms of knowledge, what is it that would have been transported?

The first thing to note about this scenario is that if the laser was going to work in its new location, the recipients would have to have a considerable background of relevant knowledge, just as Polanyi intimates. There is nothing more puzzling about this than what would happen if the mechanical bread maker were delivered to a tribe living in the Amazon jungle. To know even what it was, the natives would have to understand yeast, mechanisms, electrical sockets, and so on (once more the analogy with Stanley Kubrick's mysterious black obelisk in the film *2001: A Space Odyssey* is pertinent). In the case of the TEA laser, the recipient laboratory would have to know quite a lot about what it was and how it worked before they could make it work. If it was delivered to me, for example, I would not know how much of what sort of gas to pump through it, where to get the gas, and how to pump it in.[23] Another thing I would need to have is the counterpart of the trust in the sender mentioned above in connection with measuring the Q of sapphire (see below).

Now, suppose that the laser was plugged in and connected to the right gas supply and so forth and it worked. How is this working laser to be described in terms of knowledge transfer? The right way is to describe it in the same way one would describe a pocket calculator. Since we have established that string transformers and mechanical sequences are essentially the same thing, just as a pocket calculator is a string transformer (still more easily seen as such if we imagine it instantiated as per Babbage's Difference Engine with its brass cogs), the working laser is also a string transformer—in this case driven by analogue strings and utilizing some many-to-one mappings. The working laser, then, contains some very, very, long strings along with the means to transform them into strings that the user

23. See Collins and Kusch 1998, ch. 9, for a similar discussion regarding automated air-pumps.

can interpret. The process is the same as when I insert a shiny CD into my computer's drive and readable things appear on the screen. In the case of the laser I plug it in and feed it with gas and concrete begins to smoke; the smoking concrete is a string that affords "this laser is working" to a human observer. To this extent, it can be said to have displaced some of what was once experienced by humans as tacit knowledge with strings—explicit knowledge. The kinds of tacit knowledge that have been displaced are relational tacit knowledge and, perhaps, somatic tacit knowledge. No collective tacit knowledge has been substituted. Collective tacit knowledge is what is needed to plug it in, make it work and use it to some effect.

If explicit knowledge is strings, and longer strings can substitute for gaps in cultural proximity between sender and receiver, and the longer strings eliminate relational tacit knowledge, or a subset of somatic tacit knowledge, then it is easy to imagine that the machine "incorporates" tacit knowledge. But the machines do not have knowledge; they are just strings. In machines there is no tension with the explicit and, therefore, just as in the case of cats, dogs, and trees, the phrase "tacit knowledge" is not properly applied to machines (see pp. 72ff). And when it comes to collective tacit knowledge, there is not even any displacement or substitution going on.

There are entire philosophies built upon the incorporation of tacit knowledge into machines and things. These philosophies are just as mistaken as they ever were. The central domain of tacit knowledge—collective tacit knowledge—is beyond explication, even in sense 4 of table 4 (see p. 81), and its polimorphic aspects are beyond substitution. So the incorporation of tacit knowledge still cannot do much in the way of philosophical work. It is true that what Kuhn called a scientific or technological paradigm involves new ways of doing things with new devices and experimental apparatuses. The way of thinking is intimately tied in with the material surroundings, as, to borrow Peter Winch's extraordinarily perceptive account of the invention of a new theory, the rubber gloves, green sheets, face masks, autoclaves, sinks and disinfectants found in a hospital, are intimately related to the idea of "germ."[24] But we should not make the mistake made by Latour (see concluding section below) or that made the over-interpreters of the notion of the materially extended self. The disinfectant "contains" the idea of germs no more than my calculator "contains" arithmetic. Disinfectant is something we use because its affordances are convenient and that is also why we use the rest of the things listed in the last sentence but two, but it is all *our* tacit knowledge in an asymmetrical re-

24. Winch 1958, 121.

lationship with whatever we can make use of in the material things; the things just make our conceptual lives more convenient. Thus, when the Kuhnian idea of the scientific paradigm with its combination of concepts and practices, is used to illuminate the way science works, the proper understanding of the role of the material elements is metaphysically conservative.[25] The correct reading is along the lines that different instruments are part and parcel of different scientific lives and that a certain scientific life lived without a certain instrument would be like a surgeon who believed in germs operating in a dirty waistcoat. Nothing more mysterious should be attributed to the instruments than that they are symbolic resources with convenient affordances. Not that the use of material objects with convenient affordances is not the sign of an expert: given a choice, one would not employ a carpenter who used a heavy rock instead of a hammer even though it could still sink nails.

The incorporation of tacit knowledge does not even provide for the general theory of management such as that proposed by, say, Nonaka and Takeuchi. Both of the philosophical and the management "solutions" require that it be possible for all of tacit knowledge, rather than a subset of it, to be incorporated in material things. At the same time, the fact that some tacit knowledge can be replaced by strings in material things, explains why these philosophies have such a grip on the imagination. But machines and things don't possess knowledge; they only comprise or contain interpretable strings with certain affordances. The principle here is similar to that which applies to mimeomorphic actions. Machines can mimic mimeomorphic actions but they cannot *act*, because they do not have intentions. These differences may, nevertheless, be invisible to the outside observer.

One other missing feature of the TEA laser study is the interactional/contributory expertise dimension. If doing the same study again, it would be interesting to compare the abilities in terms of knowledge transfer of those who had merely visited other laboratories, gaining interactional expertise alone, and those who had actually worked with a laser in other laboratories, gaining contributory expertise. We know that contributory expertise is not an infallible resource. Thus, a laser builder who had actually built a working TEA laser still had trouble building a second one in a new laboratory.[26] In retrospect, this is an obvious feature of human life: accomplish-

25. For example, Galison remarks that the instruments used by high-energy physicists embody "strategies of demonstration, work relationships in the laboratory, and material and symbolic connections to the outside cultures" (1997, p. 2).

26. Collins and Harrison 1975.

ing a difficult task once might make it easier the second time but it does not remove the difficulty. Were it the case that a task done once was always easy in the future, human life would be very different. Thus, we should not expect those with contributory expertise to succeed immediately and so any difference between them and those with interactional expertise alone would be perhaps a matter of degree not kind. The problem needs further study, probably by examining, under laboratory conditions, different ways of learning some simple task.

EIGHT

Mapping the Three Phase Model of Tacit Knowledge

The aim of the book, as set out in the introduction, was to produce a Google Earth–type map of tacit knowledge. The initial condition for producing the map of tacit knowledge was to understand the concept. To understand the tacit it was first necessary to understand the explicit. Understanding the explicit was the work of Part I, and it proceeded by going back to the most basic elements—strings and things. The way strings impacted on entities was analyzed; sometimes—only when human beings were involved—the impact involved interpretation. Sometimes the impact involved only mechanical effects; humans could be affected both mechanically and via interpretation (as well as being media on which strings could be inscribed). Humans, then, can be media, can be involved in mechanical relationships with strings, or can interpret them by drawing on their collective lives. These distinctions make it possible to see the role of the body clearly and separate its contribution to what we experience as tacit knowledge from the other components. For a string to affect an entity in a way that gives it new knowledge or powers—that is, for it to have an impact that is more than mere "inscription"—more than simple transmission may be required. Sometimes a string will have to be transformed (condition 2); sometimes a longer string will work when a short string will not (condition 3); sometimes a permanent change will be required in the entity being impacted (condition 4); sometimes a change will be required in the impacted entity which creates the conditions for flexible interpretation of the string (condition 5). Condition 5 applies only to humans when interpreting.

Explicit having been understood, tacit knowledge was then broken down into three categories—RTK, STK, and CTK—and analyzed in chapters 4, 5, and 6. Going back to the main divisions, we can now draw a simple map (figure 8). Unsurprisingly, the map has three zones—an outer zone of re-

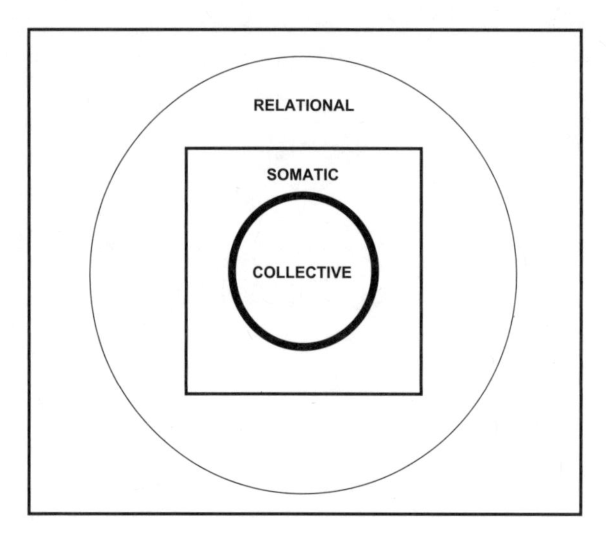

Figure 8. The terrain of tacit knowledge

lational tacit knowledge, an intermediate zone (harder to access but still attainable in principle) of somatic tacit knowledge, and a central and currently inaccessible zone of collective tacit knowledge.

Now, if we take seriously the arguments set out in the introduction, this is a very weird map because it is geographically "inside out." That is, what are presented as hard-to-access regions are in fact the regions we live in every day—the familiar world. We say that they are hard to access only because it gets harder to "tell" what goes on in them as one moves inwards. The innermost region is inaccessible in this sense by any of the means of explication listed in table 4. As the introduction argues, the inner regions of tacit knowledge are said to be hard to penetrate because the idea of tacit knowledge is formed from its tension with the idea of explicit. If figure 8 was a map of the traditional sort, which shows where we live and where we have not yet set foot, the two inner regions would be filled with familiar creatures. There would be humans in the middle living their social lives (with maybe some chimps, dolphins, and other creatures around the edge). And there would be more humans (-as-animals) along with cats, dogs, trees, sieves, and the rest of the natural world in the intermediate zone.

The outer zone of relational tacit knowledge is a bit different than the other two, because it does not contain fixed inhabitants but describes some features of human-to-human communication that, like the virtual particles that fill the vacuum according to quantum physics, are continually coming

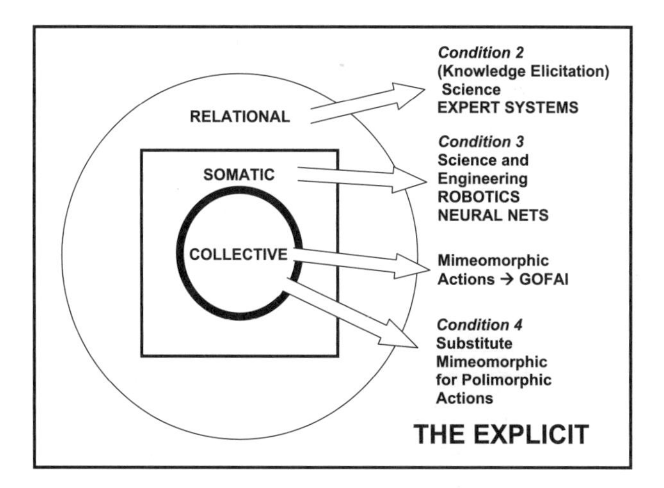

Figure 9. From tacit to explicit

into existence and going out of existence (switching between tacit and explicit), depending on the relationship of the particular humans who are in interaction. As argued in chapter 4, however, like those virtual particles, not all relational tacit knowledge can become explicit at once.

We live happily and without stress in the middle region of the map. It is the lack of stress with which we live that makes it so easy for us to confound the three types of tacit knowledge. RTK, STK, and CTK are all learned in roughly the same way—by just living in society. We enter social relationships which result in the explication of this or that piece of relational tacit knowledge, we are shown how to use our bodies to accomplish skilled tasks, and we come to absorb knowledge from the body of the social animal. All three types of development will often be needed as a competence in a specialism is gained and it is unlikely that we will notice that there are different types of increase in wisdom and competence happening at the same time. It is only when we look more carefully and with determined analytic rigor that we find that the ordinary life of tacit knowledge is a much more complicated than it seems. Familiar experiences begin to seem strange when we try to gain access to the surrounding "wilderness" of explicit knowledge. Only very close examination shows that the degree of difficulty varies because the there are three distinct reasons for the kind of difficulty that is encountered. There are three degrees of difficulty—weak, medium, and strong—associated with three kinds of obstacle. Figure 9 shows how we push our explorations into these forbidding regions of the explicit.

Working downward, relational tacit knowledge is made explicit to the extent that it can be made explicit, by telling secrets, by using longer strings, by finding out more about what is in other peoples' minds, and by doing more science so that what is not known to anyone becomes known (as in the example of the eight-inch laser lead). The "expert systems" boom of the 1980s was driven by the assumption that with enough of the kind of perseverance, known as "knowledge elicitation," the tacit knowledge of any expert could be made explicit and reproduced in a computer programming language, such as PROLOG. It is now easy to see how misguided this approach was and it is easy to understand its many failures. But the attempt to build computerized expert systems was not all failure, just as not all attempts to rectify the deficiencies of condition 1 communication with condition 3 communication are all failures—the analysis of the pub joke illustrates the point (see pp. 30–31). Of course, the idea that knowledge elicitation could render all the tacit knowledge of experts explicit was deeply wrong in a philosophical sense. First, for even the most elaborate condition 3 communication to work the recipient has to be such that the string will not bounce off or produce no more than an inscription (which is our way of saying that all explicit knowledge depends on a background of tacit knowledge). Though even here there will be occasions when failure will turn to success: some initially failing condition 3 communication can come to succeed as a result of changes in the recipient (user of an expert system)—a matter of conditions 4 and 5—but this is going to be an occasional event not a general rule. The second of the "philosophical" problems is that not all relational tacit knowledge can be made explicit at once even if any piece of it can.

Moving down, rendering somatic tacit knowledge explicit can be simply creating the condition for a formula to work in a human being where it did not work initially. In this case the condition is created by adding something to the human such as the weight-lifter's new muscles. We can then communicate the ability to carry out a "clean and jerk" merely by explaining it. Another way in which we use the term explicit in this case is when we have a scientific understanding of the working of the bodies (and brains) of humans, animals, and other things. Reproducing the effects of somatic tacit knowledge in machines and computers can also be taken to indicate that they have been understood. Hence, we say that the bicycling robots show that bicycle balancing has been understood. This is also how neural nets are sometimes thought about—they reproduce some of the abilities of human (-as-animal) and animal brains through a training program utilizing positive and negative feedback (as animals, but not humans, are typically trained).

What is meant by "reproducing human abilities," however, is not without its ambiguities. Just exactly what has been reproduced? Is it the outcome of action or is it the process of the action? In the case of chess, or in the case of the bicycling robots that utilize gyroscopes, it is easy if the criterion for understanding is taken to be simply mimicking the ability defined in some unrefined way—winning at chess and keeping the bike upright. On the other hand, if one wants to be sure that the human (or animal) mechanism has been understood then the machine has to do the job in the same way as the human. In chess, for example, it is not the case that current computers win by using human-type analysis. As argued in chapter 4, even in such cases there is no reason of principle why more scientific research might not one day enable computers to win at chess in a thoroughly humanlike way—by recognizing patterns with no more depth analysis of future moves than humans currently engage in.

There remains, however, the possibility that this might not be achieved in practice because there is something about the actual substance of the brain that makes it very difficult to reproduce in other materials; that, in other words, the human (animal) brain and body has certain affordances not shared by other materials. That even something as simple as the process of sieving is difficult to reproduce without using a sieve shows that somatic-affordance has to be taken seriously.

When we come to collective tacit knowledge the difficulties are of a different order. The exception to the intractability of collective tacit knowledge is mimeomorphic actions. Where humans act mimeomorphically, good, old-fashioned artificial intelligence (GOFAI, or symbolic AI) can succeed. At one time these small successes for symbolic AI were thought to herald the eventual mimicking of all human abilities through incremental improvements. This was never going to be the case. Collective tacit knowledge can also be said to be explicit when the receiver has enough fluency in the relevant language or social skill to be said to have benefited from an explicit communication—in this way the tacit can be rendered explicit through condition 5 changes—just making the person more fluent so that they have a better chance of benefiting from the attempted communication.

Though we can say a lot about the consequence of the existence of collective tacit knowledge and make useful predictions as a result, we do not know how it works nor the mechanisms by which individuals draw on it. Enthusiasts for neural nets believe them to be the solution to the reproduction of collective tacit knowledge but this is wrong. Where the actions are polimorphic, the only thing that can be done is to make a rough substitution with mimeomorphic actions and pretend that the wooden, routinized, non–context sensitive result, is equivalent to the rich and context-sensitive original.[1]

The history of artificial intelligence could be said to be reproduced in figure 9. Starting with GOFAI, mimeomorphic actions were reproduced and it was claimed that this was the first step to reproducing all human actions. Expert systems were then said to have "done the trick" with knowledge elicitation when it was thought that the "power is in the knowledge." Of course, what expert systems were really doing was reproducing some bits of relational tacit knowledge. Neural nets were then said to be the solution but what they do is reproduce animal conditioning regimes and certain (admittedly difficult to reproduce) aspects of somatic tacit knowledge. Nowadays we are invited to be amazed by combinations of robotics and neural nets on the one hand and the attempt to incorporate the context sensitivity and real world engagement into programs on the other. But nothing of real significance seems to be happening.

In this whole business there is a lot of misdirection going one. Deliberately or inadvertently, attention is drawn to shallow, if clever and hard to accomplish, tricks and away from the deep problems of knowledge. Pretense is all too easy because of humans' extraordinary capabilities when it comes to the repair of broken communications and, a subset of this propensity, anthropomorphism. These are the means by which a device which reproduces human action only poorly and woodenly can still be useful as a social prosthesis and therefore imagined to be a reproducing the original human activity along with its burden of collective tacit knowledge. Two other mechanisms aid the process. The first is changes in the way we live our lives so that what was once done in society via polimorphic actions becomes done mimeomorphically—and therefore can truly be reproduced by machine. An example is standardized spelling. Once standardized spelling has been introduced, machines can be used to correct it. Of course, as is argued in chapter 5 and illustrated with the argument around box 16, we have not really achieved standardized spelling, and nor do we want to, so that, in fact, our spell-checkers work crudely at best. Secondly, the ubiquity of cultural change associated with conditions 4 and 5 also make it easier for mechanisms to appear to reproduce collective tacit knowledge.

Figure 10 illustrates the way the domain of the explicit reenters the do-

1. The possibilities and limits of expert systems are examined in Collins 1990, while the meaning of mimeomorphic and polimorphic actions and the way they articulate are examined at length in Collins and Kusch 1998.

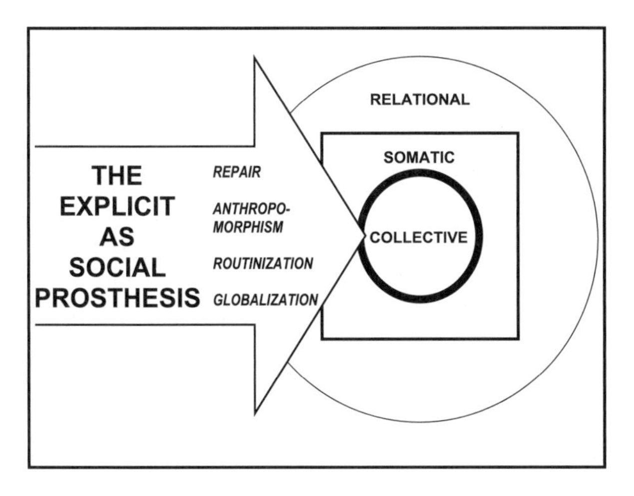

Figure 10. How machines enter social life

main of the tacit—our familiar world. Repair, anthropomorphic tendencies, perhaps some routinization of social life—nearly always less extensive than it appears to be—and, to use a term which to some extent captures the spread of changes required for condition 4 and condition 5 communications, "globalization," allows computers and other machines programmed with some transformation of the explicit to become part of the world we know and, to some extent, encourages the lazy-minded to imagine that such devices are ordinary members of our society.

How to Use the Map

A promise was made in the introduction: *The case studies and analytic discussions of tacit knowledge that we already have in hand—the bike riding, the laser building, the sapphire-quality measuring, the car driving, the natural language speaking, the breadmaking, the transfer of knowledge between organizations, and so forth—will turn out to be aspects of the territory seen from different vantage points. With the new map we, will see where those known bits of the territory are separated by mountains, where they are linked by passes, and where it was always just a matter of level ground* (p. 2). To honor the promise we need to show how to make such cases fit the map. Here is the method: take each case and split the tacit knowledge in it into RTK, STK, and CTK. Also, separate out any useful coaching and second-order rules that apply specifically to the human organism. Given this, can one see how one might go about understanding, describing, and, to the extent it can be done, automating or reproducing the abilities associated with tacit knowledge. It turns out that all the cases are similar because each spreads across the whole map. The impenetrable mountain ranges lie within each case for each case has aspects of all three elements of tacit knowledge and, as we know, the most impenetrable boundary surrounds collective tacit knowledge. In the absence of the map, the analysis of the relationship between explicit and tacit in any one case study is usually misunderstood; analysts think they have done the whole task when, in reality, they have looked at only one region. The temptation to stop before the job is finished is increased by the fact that the three kinds of tacit knowledge are indistinguishable when encountered in everyday experience.

To repeat the RTK, STK, and CTK analysis of every example in the book would be tedious. The claim made here is simply that the Three Phase Model of tacit knowledge tells the truth about how tacit knowledge is encountered in the world and applying it will enable reliable predictions to be made about what part of any activity can be reproduced mechanically and how attempts at such reproduction, or partial reproduction, should be approached.

Conclusion: The Necessity of Social Cartesianism

If the argument of this book has been convincing or useful it is worth noting the extent to which if rests on Social Cartesianism. The entire apparatus that has been deployed to explain how the tacit is turned into the explicit and how the explicit is returned to the body of social life starts with the idea that humans are special in that they have collective tacit knowledge not shared by animals and things; this gives them an asymmetrical power to organize all those things that amount to what we perceive as "selves" and to repair their deficiencies and fit them into our social lives. Once understood, the apparatus explains and can predict what the explicit can do and what it cannot do—where we can expect steady progress and where nothing will change without unforeseeable breakthroughs. Without an apparatus that turns on the special nature of human beings, we are powerless to make such predictions.

These conclusions have been drawn out piece by piece in the analysis of the different kinds of tacit knowledge. Different kinds of "cannot" have been shown to apply to different kinds of tacit knowledge—there are different reasons why we cannot make this or that piece of knowledge explicit and why we cannot build machines to mimic this or that kind of human activity. Of course, as explained in this chapter, we have mechanisms for obscuring these difficulties and that, perhaps, is why we are still promised that computers are just about to take over the world and that educational institutions and the standard methods of learning and apprenticeship can be replaced by bureaucratically organized and assessed "programs" of learning objectives. Though the main purpose of this book has been the analysis of tacit knowledge, to finish, the consequences, first theoretical and then practical, will be gathered together and filled out a little further.

Theoretical Consequences

A consequence for theories of society or knowledge is that they cannot be too symmetrical in respect of the role of humans on the one hand and animals and material things on the other. When we talk of the "extended person" or the "extended mind," we must remember that insofar as, say, a blind man's stick or an author's computer or mathematician's calculator extends their persons or minds, their hands, arms, and even the elements of their brains, such as their visual cortex, can be looked upon in the same way.[2] The point is that though it may well be that the material scaffolding in which I find myself extends "me" outwards, as far as the metaphysics of the situation is concerned, it extends me just the same way as bits of my body always have. It may be that we have discovered that the self is no longer exactly congruent with the body and brain, but the real discovery is that it never was. And the crucial implication of this is that the discovery of the extended self/mind has made no metaphysical difference to the notion of the self. The self is there just as much or as little as it ever was.

I have argued that the idea of the self cannot be eliminated because of the asymmetrical relationship between it and the things that it extends into—visual cortex, arm, hammer, stick, calculator, computer, or whatever; something has to be there to explain that asymmetry. I have made the suggestion that one clue to the asymmetry is the idea of the social; under the heading of Social Cartesianism, I have argued that the one thing that is absolutely special about humans and not about subcomponents—brains, arms, calculators, and so forth—is that they are able to learn how to make use of the contributions of all the elements in ways that act in concert *with what other humans are doing* as a result of our mutual participation in the larger organism of society.

2. The brilliant insight about the elements of the brain like the visual cortex just being part of the "extended mind" (Clark 2003) is said by Andy Clark to have come in a personal communication to him from David Chalmers (Collins, Clark, and Shrager 2008).

The same conclusion follows for theories of society or theories of knowledge that argue that human and nonhuman elements each contribute equally to the construction of the categories which organize our world. The most well-known of these approaches, at least in my area of the academic world, is Latour's actor-network theory (ANT), or as it should be called, actant-network theory, drawing as it does on the notion of "actant" found in semiotics. Latour's signature academic move was to reject the human-centered approach of the sociology of scientific knowledge (SSK) in which humans were treated as though they were largely or entirely responsible for the "social construction" of that world, including the world of scientific findings. Under SSK, the data produced by experiments were shown to be subject to sufficient "interpretative flexibility" to allow the same data to be taken to mean different things in different circumstances. Latour argued that even the very difference between humans and nonhumans was "constructed" (not "socially constructed," of course) as a consequence of the relationship of the elements (the actants) in networks; thus, knowledge-constructing networks were to be seen as metaphysically symmetrical, humans no longer having the central role.

This move, philosophically adventurous though it was, merely returned the analysis of science and technology to the position long held by philosophers and scientists in which the social and the natural each contributed to our idea of the nature of the physical world. Indeed, it may have been the very familiarity of the new/old model that contributed to its widespread acceptance in the academic field known as science and technology studies. The trouble is that the basic element of ANT, the actant is without qualities beyond those precipitated by its position in the network. This is taken to be a philosophical strength because it is more "foundational" than other theories. But it follows that humans, who are merely actants under the theory, have to be treated as symmetrical with every other actant; they have no special properties. There is, then, no room for old-fashioned Cartesian duality under ANT, an attractive feature for many modern thinkers, but there is no room for Social Cartesianism either. This renders the approach impotent when it comes to explaining many features of the relationship between humans and things—the more applied consequences of Social Cartesianism that have been mentioned above and that will be set out again below.[3]

In the well-known moral tale, a sage is explaining that Atlas supports the world. A member of the audience asks what supports Atlas, to which the

3. Many of the criticisms can be first found in Collins and Yearley 1992.

sage replies "an elephant." "And what supports the elephant?" comes the response. "A turtle," says the sage, "and from there on it's turtles all the way down." The moral we ought to draw from this is that no argument can ever be justified on the sole grounds that it takes us to a lower turtle—one can go on doing that indefinitely—the justification has to be in terms of why it is a better turtle.

Too much symmetry is also a problem that applies to any theory that works toward eliminating the boundary between humans and animals. It may well be true that the boundary is fuzzy, but that, as has been argued above, is not the same as saying there is no distinction to be made. Animals are simply not embedded in human society in the way humans are embedded and insofar as there are animal societies they are not even remotely as rich as human societies, where richness is indicated by the amount of cultural variation within any one species. To see what culture means, look not for the accomplishment but for the diversity. To repeat, there are no dogs that are vegetarian on principle, nor will there ever be unless we find out how to change dogs so that they can be socialized.

Ironically, in this book humans have been closely compared to animals; it has been claimed that somatic-limit tacit knowledge is ontologically on a par with the bodily abilities of animals. Dogs can ride bicycles and could probably be trained to drive cars—but not in traffic. Animals could mimic the full range of mimeomorphic actions. (We have to say "mimic" because the notion of intention, which is a defining part of actions, is not applicable to animals in the same way). But as the book points out, it is those very things that the animals cannot do—polimorphic actions informed by the collective tacit knowledge—that renders humans unique. Indeed, one might mildly criticize phenomenological treatments of the body along the lines that much of what they say about humans goes for animals too, and that is why they miss the special qualities of specifically human experience. Consider the notion of "body schema"[4]: it is the body's relationship with the world that gives rise to our understandings of it. But surely a similar relationship must exist between the bodies of dogs or lions and the world; their relationship with the world will be different from species to species, as well as between each species and the human species, because the bodies are different. Still, it seems odd to say that dogs, lions, and humans "understand" the world differently. This is because dogs and lions never try to make their image of the world explicit so we do not worry about what they

4. Gallagher 2005.

think. Wittgenstein remarked, "if a lion could speak we would not understand him,"[5] but the unique feature of humans is that they do speak and that is why we are caused to think about the relationship of our bodies to the structure of our concepts.

"Kinesthesia" is "an awareness of the spatio-temporal and energy dynamics of one's movement." It has been argued that "An agent devoid of kinesthesia . . . belongs to no known natural species." "Agents—those having the power to act—necessarily have a kinesthetic sense of their own movement."[6] This leads to a fascinating discussion by Gallagher of a person with no sense of touch and no proprioception below the neck but who can still move and act normally. Gallagher argues that the normal movement can be explained by the residual kinesthesia from the neck upward, particularly that associated with eye movements.[7] And yet one might still have the same discussion about an injured dog but it would still miss the essential feature of humans which is that we can talk of the abilities that follow from kinesthesia as "tacit knowledge" only because humans also have explicit knowledge. In other words, this kind of discussion, fascinating and important as it is for questions to do with human and animal experience, does not bear upon the special quality of human action.

The fact that we can speak also has the previously overlooked consequence that in the case of humans, though *not* in the case of animals, the structure of our concepts is formed at the collective level so that the conceptual structure of someone born with no legs is the same as that of someone with legs; that conceptual structure comes from the language, not the individual body's interactions with the environment. In the case of animals, being born without legs (in the wild) is a death warrant—a rabbit without legs is not a rabbit except insofar as it shares its DNA with other rabbits; in the case of humans, being born without legs is a physical inconvenience and in terms of concepts it is hardly noticeable. That is a huge difference between animals and humans: to repeat, in the one case the body has its influence on the individual, in the other case the body has its influence almost entirely at the collective level and this is all to do with language. Language is the link between the collective body and the individual. Language is like DNA—language, like DNA, identifies the human as a human irrespective of the shape of the individual's body. But language can also discriminate at a lower level. It is the language that is the identifying feature

5. Wittgenstein 1967, 223 (see also box 17 on p. 134).
6. Sheets-Johnstone, quoted in Gallagher 2005, 61.
7. Ibid., 61ff.

of a specific social group of humans even though that group has DNA, and a collective body shape, in common with all other humans. In the case of both ANT and too enthusiastic a dissolution of the boundary between humans and animals, the mistake is the symmetry—it is humans alone who have this linguistic equivalent of DNA; we cannot understand the world without the asymmetry of Social Cartesianism.

Practical Consequences

Many of the practical consequences of the argument of this book have been long anticipated. In the face of vicious opposition, Hubert Dreyfus pioneered the broad approach in 1972 and has remained in the van ever since, exemplified by his 2001 book *On the Internet*. The only things Dreyfus got wrong were the underlying theory and the details. His mistake, based on his belief that Heidegger was the greatest philosopher of the twentieth century,[8] was to take the nature of the body as the fundamental barrier to the indefinite extension of the "intelligence" of computers. It all began with his 1965 paper entitled "Why Computers Must Have Bodies in Order to Be Intelligent." But a much better case can be made for Wittgenstein (the later incarnation) being the key thinker, with the nature of social life giving rise to the fundamental limits. The reason Dreyfus was such a deservedly successful pioneer, aside from his academic heroism, is that there is a big overlap between what can be concluded about our relationship to computers irrespective of whether it is the body or the society that is taken as fundamental. But the overlap is not complete and the divergence between the two approaches is likely to grow as the domain of the body, STK, is more and more invaded by a combination of robotics and neural nets.

Taking the social as the key, one must look in two opposite directions. The first direction is to the ways in which societies change; the very cultural flexibility open to humans allows them to change their societies in ways that can render the impossible possible. Thus, the Dreyfusian five-stage model suffers from its failure to distinguish gear shifting from traffic scenario recognition which, as it was argued above, explains why there are automatic gearboxes but no automatic scenarioboxes. But it would be possible to make the equivalent of automatic scenarioboxes by rearranging the world. If every car, human, dog, and cat was fitted up with responder that could be read by satellite, and nobody minded changing the car from an instrument of gratification to a pure instrument of transport, it would

8. Dreyfus, personal conversation, October 10, 2008.

be conceptually easy to automate the entire journey to work while avoiding running over humans, cats, and dogs. We would simply have substituted mimeomorphic procedures for polimorphic actions as far as driving was concerned. Once more, Orwell's *1984* intimates how something of the sort might be approached even in the domain of language; it is a matter of the extinction of troublesome cultural diversity and context sensitivity. In smaller ways, this kind of thing is happening all around us.

The other direction in which the idea of the social causes us to look is toward the nature of society as we have it. Society, which is responsible for all our collective tacit knowledge, runs on socialization and language. What this means is that education cannot be broadcast, only information can be broadcast. The fifth enabling condition of communication cannot be replaced wholesale with the fourth enabling condition of communication. Education is socialization—it the common learning of a language. Some of it can be substituted by information transmission but when too much is substituted it becomes something else. Education, then, is always going to be inefficient. And, since the Internet is a broadcast medium, it is not going to substitute for education.

The dream that the carbon footprint of business travel in airplanes will be eliminated by teleconferencing is just a dream. Much of what happens in business meetings or in scientific conferences is mutual socialization and mutual learning of language; it is the acquisition of the interactional expertise needed for the coordination of polimorphic actions in collections of humans. Again, some of this can be substituted by teleconferencing, but not all of it. You cannot build or acquire CTK with television.

Businesses that want to make money from other businesses' ideas are still going to have to buy the business or buy the employees rather than steal the plans or formulas. The locus of the crucial part of knowledge, CTK, is people and their living language, not strings.

Barring some unforeseeable breakthrough involving the solution of the socialization problem, there will never be a spell-checker that could check the spelling of this manuscript in the same way as a competent human editor. It's not that the machine won't get the correctly spelled words right, it is that the pesky context means that sometimes words are rightly speeled wrong. This does not mean that spell-checkers are not immensely useful—all automated machines are useful so long as they are helping the human check that they have made no mistakes rather than putting the mistakes right. The problem of building an automated humanlike speech transcriber is still harder because it extends even to the correctly spelled words. Whereas strings produced by a keyboard are combinations of about two hundred unambiguous elements, spoken strings are usually so badly formed that no amount of information processing can put them right. The human speech-transcriber falls back on meaning but the machine cannot. So the choice is to speak like a machine to a greater or lesser extent, or to spend a lot of time correcting mistakes.

Finally, when it comes to the nitty-gritty of making machines that do any of the jobs of humans the right place to start is a careful and detailed analysis of the task in terms of RTK, STK, and CTK and an analysis of whether the targeted users of the machine are likely to meet the conditions of communication required for the strings to have the desired effect. The scope and recipe for automation can then be found in figure 9. RTK implies knowledge elicitation, expert systems and so on, STK implies robotics, or neural nets and the like, while the impossible bit is always going to be CTK—that is where human beings have to be left to get on with it or mimeomorphic actions have to be substituted for polimorphic actions. The tasks can then be broken down still further into their component actions with each element of mimeomorphicity and polimorphicity separated out (see appendix 1). As can be seen, now that the human race has explicit knowledge, the domains in which an understanding of tacit knowledge is needed run from high theory, through education, to the design of the everyday devices.

APPENDIX 1

An "Action Survey"

Table A-1 exemplifies what can be called an "action survey." It describes an automated domestic bread maker of the type discussed by Nonaka and Takeuchi. The analysis is adapted from a previously published article.[1] Inevitably, the table shows just part of the entire "action tree" of breadmaking with a machine. The tree extends out into the rest of the world via the design of bread makers, the manufacture and selling of bread makers, the supplying of electricity to bread makers, the repairing of bread makers, the serving and eating of bread—it ranges upward and downward indefinitely.[2] Again, to get a sense of the range, imagine the bread maker parachuted to an Amazon tribe.

The action survey shows that some actions are done only by master bakers and some only by machine users. (Actually, choosing the level of tolerance of the product is done by the machine's designers, not the users.) Many of the actions have the same "morphicity," whether done by a master baker or a human using a machine. Mixing and kneading is the locus of the action tree on which Nonaka and Takeuchi concentrate their analysis. But, as explained in chapter 7, this turns out to be largely RTK. It is RTK insofar as strings that can be interpreted in such a way as to ease the building of a mechanical kneading device can, with enough effort, be "elicited" from the master baker. STK is required for a human to knead bread. Many of the actions do have to be changed quite a bit for automation to work, however. For instance, "picking up the ingredients" is likely to be a very different action for the baker than for the machine user , and there is far less choice of

1. Ribeiro and Collins 2007.

2. For morphicity and the idea of action trees, see Collins and Kusch 1998.

Table A-1. Part of an action tree involving an automated bread maker.

		BREAD-MAKING MACHINE (usually with restricted choice)	
ACTIONS	MASTER BAKER	Mimicked by machine	Substituted by humans
Setting up the production scene	*Polimorphic*		*Polimorphic*
Choosing recipe, size and crust color	*Polimorphic*		*Polimorphic*
Dealing with the variability of ingredients and brands	*Polimorphic*		*Polimorphic*
Choosing level of tolerance of final product	*Polimorphic*		*Polimorphic*
Picking up the ingredients	*Mimeomorphic*		*Mimeomorphic*
Measuring	*Polimorphic*		*Mimeomorphic*
Setting program	—		*Mimeomorphic*
Setting dough size	—		*Mimeomorphic*
Mixing and kneading	*Mimeomorphic*	*Mimeomorphic*	
Shaping	*Mimeomorphic*	*Mimeomorphic*	
Baking	*Mimeomorphic*	*Mimeomorphic*	

types of bread. But the choice of the restricted range of types made available by the machine is still a matter of polimorphicity—it depends on the social context of the activity; this aspect of things would cease to be polimorphic only if the machine offered no choice, or if the choice was random, or if it followed a preprogrammed sequence of variations so the machine "chose" the type. The setting of the program that delivers the chosen type is mimeomorphic—it involves pushing a button that is always executed with what amounts to the "same behaviors." The one place where the morphicity of an action has to be changed is measuring ingredients; here the baker makes judgements by eye according to the exact taste required, adding, for example, a pinch of salt more or less—a context-sensitive action—whereas the machine user uses exact, context-insensitive measurements from which, by providing measuring cups and instructions for leveling these off without too much tamping down, the manufacturer tries to remove as much STK as possible. One can see from the table where one must concentrate automation resources to increase the gradations of available human choice, or where to remove the responsibility for choice, or where to eliminate tedious work, and where one is forced to change the action morphicity from polimorphic to mimeomorphic, thus changing society, in a more or less significant way, if success is to be achieved.

APPENDIX 2

What Has Changed since the 1970s

In the introduction, it was claimed that the development of intelligent machines and the close study of scientific knowledge that was begun in the early 1970s has led to a deeper understanding of knowledge. These are some of the things that have changed since the early days and that have fed into the analysis of this book:

There is no knowledge barrier

The difference between computers and humans does not lie in the difference between science and mathematics versus less structured knowledge. We know this because we now know that even science and mathematics are tacit knowledge–laden. The difference is something much more subtle and fine-grained that can be unpicked with the Three Phase Model of tacit knowledge and the idea of action morphicity. This realization has come out of the close analyses of science carried out over the last three decades.

Social prostheses

The idea of the social prosthesis arises out of noticing that machines that cannot accomplish the things that humans can accomplish are often treated as though they are equivalent. The mechanism is "repair" of the machines' broken output. The key ingredient in this analysis is the success of machines in doing calculations and so forth, contrasted with the close analysis of science, which makes it puzzling that they work at all, because even science is a quintessentially social activity.

Mimeomorphic actions

The idea of mimeomorphic actions also arises out of the undeniable success of computers, even when repair is discounted. The idea of mimeomorphic actions accounts for this success by drawing on the fact that sometimes humans choose to act like machines. Polimorphic actions are the counterpart of mimeomorphic actions.

Interactional expertise

That human individuals with bodies very different than the norm can acquire knowledge that is not dissimilar to those with typical bodies has something to do with the importance of language in human societies. Natural languages are themselves tacit knowledge–laden. This arises as a by-product of close studies of science.

Social and minimal embodiment

Closely related to the last point is the discovery (which is obvious once it is stated) that much of the potential of individuals is provided by the society in which they are embedded rather than by their individual bodily form, even though the shape of the body pertaining to the species as a whole affects the nature of the society in deep ways. That human individuals with uncharacteristic bodies can share social knowledge follows from the fact that we are social parasites.

REFERENCES

Adams, D. 1979. *The Hitchhiker's Guide to the Galaxy*. London: Pan Books.

Barrow, John, D. 1999. *Impossibility: The Limits of Science and the Science of Limits*. New York: Oxford University Press.

Baumard, Philippe. 1999. *Tacit Knowledge in Organizations*. London: Sage.

Benner, Patricia. 1984. *From Novice to Expert: Excellence and Power in Clinical Nursing Practice*. London: Addison Wesley.

Bijker, W., T. Hughes, and T. Pinch, eds. 1987. *The Social Construction of Technological Systems*. Cambridge, MA: MIT Press.

Block, N. 1981. Psychologism and Behaviourism. *The Philosophical Review* XC:5–43.

Bloor, David 1999. Anti-Latour. *Studies in History and Philosophy of Science* 30:1, 81–112.

Callon, Michel. 1986. Some Elements of a Sociology of Translation: Domestication of the Scallops and the Fishermen of St Brieuc Bay. In *Power, Action and Belief: a New Sociology of Knowledge?* ed. J. Law, 196–233. London: Routledge & Kegan Paul.

Callon, Michel. 2008. Economic Markets and the Rise of Interactive *Agencements*: From Prosthetic Agencies to "Habilitated" Agencies. In *Living in a Material World,* ed. Trevor Pinch and Richard Swedberg, 29–56. Cambridge, MA: MIT Press.

Chandler, Daniel. 2001. *Semiotics: The Basics*. Abingdon, UK: Routledge.

Cho, Adrian. 2007. Program Proves That Checkers, Perfectly Played, Is a No-Win Situation. *Science* 317, no. 5836 (July 20, 2007): 308–9.

Clark, Andy. 2003. *Natural-Born Cyborgs: Minds, Technologies and the Future of Human Intelligence*. New York: Oxford University Press.

Collins, H. M. 1974. The TEA Set: Tacit Knowledge and Scientific Networks. *Science Studies* 4:165–86.

Collins, H. M. 1985. *Changing Order: Replication and Induction in Scientific Practice*. Chicago: 2nd ed., University of Chicago Press, 1992.

Collins, H. M. 1990. *Artificial Experts: Social Knowledge and Intelligent Machines*. Cambridge, MA: MIT Press.

Collins, H. M. 1998. Socialness and the Undersocialised Conception of Society. *Science, Technology and Human Values* 23:494–516.

Collins, H. M. 2001a. Tacit Knowledge, Trust, and the Q of Sapphire. *Social Studies of Science* 31:71–85.

Collins, H. M. 2001b. What Is Tacit Knowledge? In *The Practice Turn in Contemporary*

Theory, ed. Theodore R. Schatzki, Karin Knorr Cetina, and Eike von Savigny, 107–19. London: Routledge.

Collins, H. M., 2004. *Gravity's Shadow: The Search for Gravitational Waves*. Chicago: University of Chicago Press.

Collins, H. M. 2007a. Bicycling on the Moon: Collective Tacit Knowledge and Somatic-Limit Tacit Knowledge. *Organization Studies* 28:257–62.

Collins, H. M., ed. 2007b. Case Studies in Expertise and Experience. Special issue, *Studies in History and Philosophy of Science* 38, no. 4.

Collins, H. M., Andy Clark, and Jeff Shrager. 2008. Keeping the Collectivity in Mind? *Phenomenology and the Cognitive Sciences* 7:353–74.

Collins, H. M., and Robert Evans. 2007. *Rethinking Expertise*. Chicago: University of Chicago Press.

Collins, H. M., Robert Evans, and Michael Gorman. 2007. "Trading Zones and Interactional Expertise." Case Studies of Expertise and Experience. Special issue, *Studies in History and Philosophy of Science* 38, no. 4: 657–66.

Collins, H. M., Robert Evans, Rodrigo Ribeiro, and Martin Hall. 2006. Experiments with Interactional Expertise. *Studies in History and Philosophy of Science* 37:656–74.

Collins, H. M., and R. Harrison. 1975. Building a TEA Laser: The Caprices of Communication. *Social Studies of Science* 5:441–50.

Collins, H. M., and M. Kusch. 1998. *The Shape of Actions: What Humans and Machines Can Do*. Cambridge, MA: MIT Press.

Collins, H. M., and Gary Sanders. 2007. "They Give You the Keys and Say 'Drive It': Managers, Referred Expertise, and Other Expertises." Case Studies of Expertise and Experience. Special issue, *Studies in History and Philosophy of Science* 38, no. 4: 621–41.

Collins, H. M., and Steven Yearley. 1992. Epistemological Chicken. In *Science as Practice and Culture,* ed. A. Pickering, 301–26. Chicago: University of Chicago Press.

Crist, Eileen, 2004. Can an Insect Speak? The Case of the Honeybee Dance Language. *Social Studies of Science* 34: 1, 7–43.

Dewdney, A. K. 1984. On the Spaghetti Computer and Other Analog Gadgets for Problem Solving. *Scientific American* 250, no. 6: 15–19.

Dreyfus, Hubert L. 1965. Why Computers Must Have Bodies in Order to Be Intelligent. *Review of Metaphysics* 21:13–32.

Dreyfus, Hubert L. 1972. *What Computers Can't Do*. New York: Harper and Row.

Dreyfus, Hubert L. 1992. *What Computers Still Can't Do*. Cambridge, MA: MIT Press.

Dreyfus, Hubert L. 1996. Response to My Critics. *Artificial Intelligence* 80:171–91.

Dreyfus, Hubert L. 2001. *On the Internet*. London: Routledge.

Dreyfus, Hubert L., and Stuart. E. Dreyfus. 1986. *Mind Over Machine: The Power of Human Intuition and Expertise in the Era of the Computer*. New York: Free Press.

Durkheim, Emile. 1933. *The Division of Labor in Society*. Glencoe, IL: Free Press.

Dutta, Shantanu, and Allen M. Weiss. 1997. The Relationship between a Firm's Level of Technological Innovativeness and Its Pattern of Partnership Agreements. *Management Science* 43, no. 3: 343–56.

Galison, P. 1997. *Image and Logic: A Material Culture of Microphysics*. Chicago: University of Chicago Press.

Gallagher, Shaun. 2005. *How the Body Shapes the Mind,* Oxford: Oxford University Press.

Gibson, James, J. 1979. *The Ecological Approach to Visual Perception*. Mahwah, NJ: Lawrence Erlbaum.

Giere, Ronald. 2006. *Scientific Perspectivism*. Chicago: University of Chicago Press.

Gladwell, Malcolm. 2005. *Blink: The Power of Thinking without Thinking*. London: Allen Lane.

Gourlay, Stephen N. 2004. Knowing as Semiosis: Steps Towards a Reconceptualization of "tacit knowledge." In *Organizations as Knowledge Systems*, ed. H. Tsoukas and N. Mylonopoulos. London: Palgrave Macmillan.

Gourlay, Stephen N. 2006a. Conceptualizing Knowledge Creation: A Critique of Nonaka's Theory. *Journal of Management Studies* 43, no. 7: 1415–36.

Gourlay, Stephen. N. 2006b. Towards Conceptual Clarity concerning "Tacit Knowledge": A Review of Empirical Studies. *Knowledge Management Research and Practice* 4, no. 1: 60–69.

Grene, Marjorie, ed. 1969. *Knowing and Being: Essays by Michael Polanyi*. London. Routledge & Kegan Paul.

Haraway, Donna J. 2003. *The Companion Species Manifesto: Dogs, People, and Significant Otherness*. Chicago: Prickly Paradigm Press.

Haugeland, J. 1985. *Artificial Intelligence: The Very Idea*. Cambridge, MA: MIT Press.

Hedesstrom, T., and E. A. Whitley. 2000. What Is Meant by Tacit Knowledge: Towards a Better Understanding of the Shape of Actions. In *8th European Conference on Information Systems*, pp. 46–51, Vienna.

Heller, Joseph. 1961. *Catch 22*. Repr. ed., London: Vintage, 1994.

Herbig, Britta, Andre Bussing, and Thomas Ewart. 2001. The Role of Tacit Knowledge in the Work Context of Nursing *Journal of Advanced Nursing* 34, no. 5: 687–95.

Hutchins, Edwin. 1995. *Cognition in the Wild*. Cambridge, MA: MIT Press.

Kenneally, Christine. 2008. So You Think You're Unique. *New Scientist* May 24, 29–34.

Knorr Cetina, Karin. 1999. *Epistemic Cultures: How the Sciences Make Knowledge*. Cambridge MA: Harvard University Press.

Kuhn, Thomas S. 1961. The Function of Measurement in Modern Physical Science. *Isis* 52:162–76.

Kuhn, Thomas S. 1962. *The Structure of Scientific Revolutions*. Chicago: University of Chicago Press.

Kusch, Martin. 2002. *Knowledge by Agreement: The Programme of Communitarian Epistemology*. Oxford: Oxford University Press.

Latour, Bruno. 2005. *Reassembling the Social: An Introduction to Actor-Network-Theory*. Oxford: Oxford University Press.

Latour, Bruno, and Steve Woolgar. 1979. *Laboratory Life: The Social Construction of Scientific Facts*. London and Beverly Hills, CA: Sage.

MacKenzie, Donald. 2001. *Mechanizing Proof: Computing, Risk, and Trust*. Cambridge, MA, and London: MIT Press.

Mackenzie, Donald. 2008. Producing Accounts: Finitism, Technology and Rule-Following. In *Knowledge as Social Order: Rethinking the Sociology of Barry Barnes*, ed. Massimo Mazzotti, 99–117. London: Ashgate.

McConnell, J. V. 1962. Memory Transfer Through Cannibalism in Planarians, *Journal of Neurophysiology* 3:42–48.

Norman, D. N. 1988. *The Design of Everyday Things*. New York: Doubleday.

Pinch, T. J. 1982. Kuhn—The Conservative and the Radical Interpretations. *4S Newsletter* 7, no. 1: 10–25. Repr. *Social Studies of Science* 27 (1997): 465–82.

Pinch, T., H. M. Collins, and L. Carbone. 1996. Inside Knowledge: Second Order Measures of Skill. *Sociological Review* 44:163–86.

Polanyi, Michael. 1958. *Personal Knowledge*. London: Routledge & Kegan Paul.

Polanyi, Michael. 1966. *The Tacit Dimension*. London: Routledge & Kegan Paul.

Polanyi, Michael. 1969. The Logic of Tacit Inference. In *Knowing and Being: Essays by Michael Polanyi*, ed. Marjorie Grene, 140–44. Chicago: University of Chicago Press.

Polanyi, Michael, and Harry Prosch. 1975. *Meaning*. Chicago: University of Chicago Press.

Powers, Richard. 2004. *Galatea 2.2*. New York: Picador.

Ribeiro, R. 2007. The Language Barrier as an Aid to Communication. *Social Studies of Science* 37, no. 4: 561–84.

Ribeiro, R., and H. M. Collins. 2007. The Bread-Making Machine, Tacit Knowledge, and Two Types of Action. *Organization Studies* 28, no. 9: 1417–33.

Sacks, O. 1985. *The Man Who Mistook his Wife for a Hat*. London: Duckworth.

Schatzki, Theodore R. 2003. A New Societist Social Ontology. *Philosophy of the Social Sciences* 33, no. 2: 174–202.

Schatzki, Theodore R. 2005. The Sites of Organizations. *Organization Studies* 26:465–84.

Schilhab, Theresa. 2007. Interactional Expertise through the Looking Glass: A Peek at Mirror Neurons. *Studies in History and Philosophy of Science* 38, no. 4: 741–47.

Searle, John R. 1969. *Speech Acts: An Essay in the Philosophy of Language*. Cambridge: Cambridge University Press.

Searle, John R. 1980. Minds, Brains, and Programs. *Behavioural and Brain Sciences* 3: 417–24.

Selinger, Evan, Hubert L. Dreyfus, and H. M. Collins. 2007. "Embodiment and Interactional Expertise." Case Studies of Expertise and Experience. Special issue, *Studies in History and Philosophy of Science* 38, no. 4: 722–40.

Shannon, C. E. 1948. A Mathematical Theory of Communication. *Bell System Technical Journal* 27: 379–423; 623–56.

Sheets-Johnstone, Maxine. 1998. Consciousness: A Natural History. *Journal of Consciousness Studies* 5, no. 3: 260–94.

Simon, H. A. 1969. *The Sciences of the Artificial*. Cambridge, MA: MIT Press.

Simon, H. A. 1973. The Structure of Ill-Structured Problems. *Artificial Intelligence* 4:181–281.

Tsoukas, Haridimos. 2005. *Complex Knowledge*. Oxford: Oxford University Press.

Wells, H. G. 1904. The Country of the Blind. Repr. in *The Complete Short Stories of H. G. Wells*, ed. John Hammond, 846–70. London: Phoenix Press, 1998.

Whitley, E. A. 2000. Tacit and Explicit Knowledge: Conceptual Confusion around the Commodification of Knowledge. In *Knowledge Management: Concepts and Controversies*, ed. H. Scarbrough and R. Dale, 62–64. Business Process Resource Centre, Warwick University.

Winch, Peter. 1958. *The Idea of a Social Science*. London: Routledge & Kegan Paul

Wittgenstein, L. 1953. *Philosophical Investigations*. Oxford: Blackwell.

INDEX